EXPERIMENTS FOR

ELECTRICAL MACHINES, DRIVES,

AND POWER SYSTEMS

**for Electrical Engineering
Technology Students**

Stephen P. Tubbs
formerly of the
Pennsylvania State University
currently
working in industry

Third Edition

CONTENTS

iv

PART IV MOTOR CONTROL EXPERIMENTS

     Relay-Contactor Control

     Power Electronic Control

INTRODUCTION

Electric motors, transformers, and control systems are used in all modern industries. A knowledge of the characteristics of these will help the electrical engineering technologist keep the wheels of industry turning.

This book will give the student a practical introduction to electrical machinery, transformers, and motor control.

The experiments have all been used at the Pennsylvania State University, Mckeesport Campus. There, the full series of experiments were done in two semesters. Each experiment requires about two hours of laboratory time.

The book is designed to accompany a textbook.

As an added feature, the book also has sections on conducting an experiment, laboratory report writing, accuracy, equipment, and motor runaway.

It should be understood that not all of the included experiments will be possible at all schools. Different schools have different equipment strengths. Some of this book's experiments will serve only as a guide for instructors setting up experiments with their own equipment.

Stephen P. Tubbs

CONDUCTING AN EXPERIMENT

Safety is a primary concern in conducting an experiment.

SAFETY RULES

1. Do not engage in horseplay.
2. Use good leads and repair or replace leads with loose connectors or poor insulation. Make solid connections, because poorly made connections may be shaken apart by motor vibrations.
3. Avoid touching rotating parts or allowing wires to touch them.
4. Double check every circuit before energizing it.
5. Make certain that all members of your group realize when a circuit is being energized.
6. Remember that voltages used in this laboratory are high enough for electrocution.
7. When in doubt, ask the instructor.
8. Use good sense and be careful.

WHAT TO DO IN CASE OF A SERIOUS ACCIDENT

1. Turn off the equipment completely.
2. Examine the victim.
   a. If the victim's breathing has stopped, as could be the case with electrocution, start mouth-to-mouth resuscitation.
   b. If the victim's heart has stopped beating, as could be the case with electrocution, start cardiopulmonary resuscitation.
   c. If there is serious bleeding, apply direct pressure to the wound.
3. Call an ambulance.

SUGGESTIONS ON HOW TO CONDUCT A SUCCESSFUL EXPERIMENT

1. Read the experiment thoroughly before the start of the laboratory period. Usually, it is best to read it the night before to allow the experiment time to "sink in".
2. At the start of the laboratory period, identify the equipment and record the equipment's nameplate data and identifying numbers. Nameplate data for equipment are particularly useful because they let the user know how much current and voltage the equipment should receive.
3. Divide the construction of circuits up among the different people in your laboratory team. One person can connect the first circuit, another the second, and so on.

4. Construct each circuit neatly. Use color-coded leads if possible and use the shortest leads that will accomplish the task. In some experiments putting paper labels on terminals and meters will make construction and operation easier.

5. Be certain that all connections are good and not likely to shake apart.

6. Have each member of your laboratory team check the circuit. When the laboratory procedure requests it, have the instructor check the circuit.

7. Make tables for recording experimental data like those of Experiments 2 and 6. Neatness in recording data will pay off later when the laboratory report is written.

8. When possible, it is best to energize the circuits gradually. Often circuits are powered by variable voltage supplies. Those circuits can first be energized at low voltage and then brought up to operating voltage. As the voltage is brought up, the experimenters would watch for any unusual or incorrect meter reading, unusual noise, or smoke. Of course, not all circuits should be energized this way, but most electrical machinery and transformer circuits can be without damage.

9. Divide up the taking of readings among the different people on the laboratory team. Usually it is best to designate one person as the laboratory team foreperson, so that he or she can tell the others when to read their assigned meters.

10. Be considerate to the next people who will use the equipment. Stay alert for smoke, burning smells, or unusual noises coming from the equipment. When in doubt, turn off the power and ask the instructor.

11. Before disassembling a circuit to go onto the next, carefully consider the data recorded. Do the data agree with what was expected? If the data seem questionable, double check the circuits, meter scales and a few of the readings.

12. When the experiment is finished, neatly put away the equipment.

LABORATORY REPORT WRITING

A laboratory report should be written for each experiment, unless it is indicated otherwise.

RECOMMENDED LABORATORY REPORT FORMAT

I. Objective: A statement of the objective of the experiment.

II. Materials: A list of all major pieces of equipment used. Equipment listed is identified by serial numbers to allow duplication of the experiment, if necessary. Particular attention is paid to recording the complete data of the electrical machine being tested. Nameplate data are often useful in answering questions asked in the question section and are themselves instructive.

III. Procedure: The procedure used in the laboratory experiment reported in the student's own words.

IV. Questions: Answers to the questions.

V. Conclusions: Summary of the important concepts learned from the experiment.

LABORATORY REPORT WRITING GUIDELINES

1. Be neat.
    a. Use 1 1/4" margins for all work.
    b. Do all writing, drawings, tables, and graphs so that they are upright without turning the sheet. The only exceptions to this rule will be a few of the data tables.
    c. Use CAD or drafting instruments for circuit diagrams. Do drafted drawings in pencil.
    d. Make data tables like those of Experiments 2 and 6. Each data table should have a title and have its appropriate quantities and units at the top of each column. Data may be written in pencil.
    e. Prepare graphs on high quality graph paper as shown in the example graph of Figure 1. There should be a title above the graph. Each axis should be labeled with a quantity and its units. The graphed lines should be straight lines or smooth curves that come close to but do not necessarily go through the data points. There should be a descriptive label next to each curve; the label should be more than just a procedure number.

4

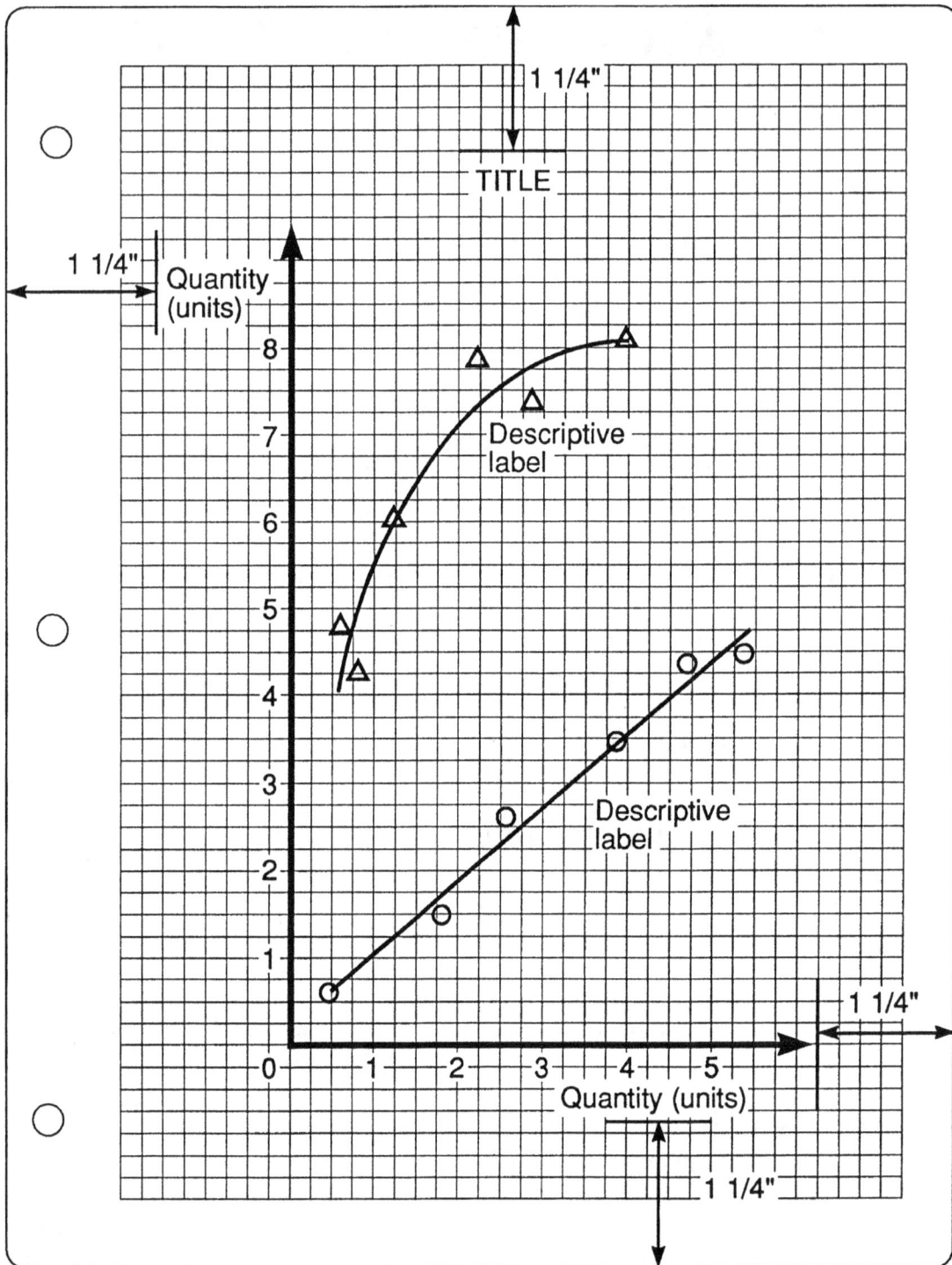

FIGURE 1

f. Word process the main body of each report or write it in ink. Word-processing isn't required, but it does improve the clarity of some reports. Use your judgment to decide if it is worth it.

g. Number answers with the same numbers as those of their corresponding questions.

2. Be complete.
    a. Answer questions completely.
    b. Use complete sentences.
    c. Show calculations. (If a calculation is repeated often, an example calculation is sufficient.)

3. Be correct.
    a. Answer questions correctly.
    b. Use correct grammar.
    c. Use correct spelling.

4. Write in the third person impersonal. Avoid using I, me, mine, you, your, he, him, his, she, her, hers, we, us our, they, them, their('s).

5. Remember that a laboratory report's worth is proportional to the effort that went into writing it.

PART I

DC GENERATOR AND MOTOR

EXPERIMENTS

EXPERIMENT 1

MOTOR PARTS

-OBJECTIVE-

To identify common electric motor parts.

-EQUIPMENT-

Dc or universal motor armature
Motor stator (dc or ac)
Dc or universal motor armature lamination
Motor stator lamination

-PROCEDURE-

1. Match the parts shown in the drawings with the answers given and with the actual motor parts supplied.
2. No report is required for this experiment.

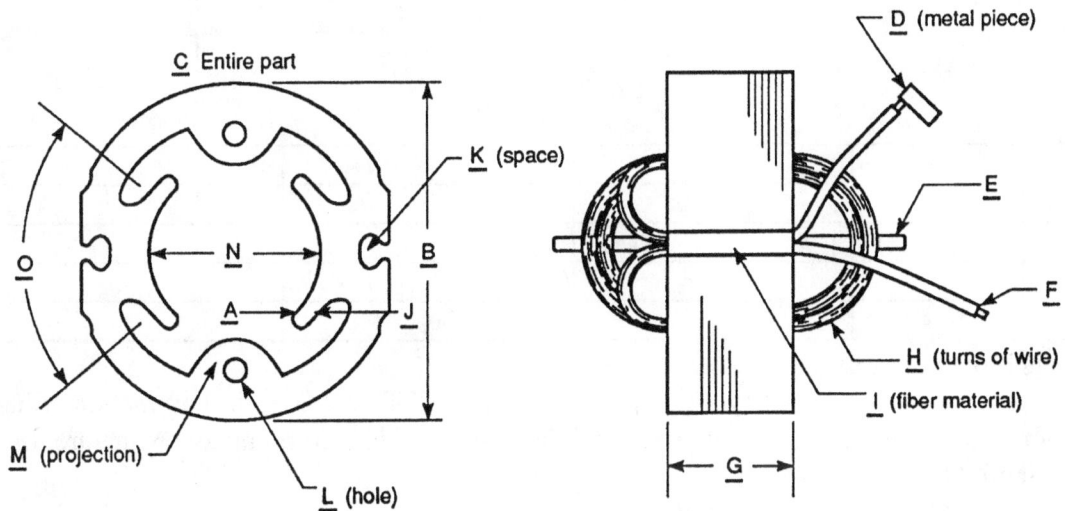

FIGURE 1-1

A	E	I	M
B	F	J	N
C	G	K	O
D	H	L	P

Answers (not in order):

Stack Height, Wedge Slot, Lead, Clamp Bolt Hole, O.D., Bore or I.D., Pole Face, Clamp Bolt Boss, Lead Terminal, Winding Pin, Span, Stator Lamination, Slot Wedge, Stator Coil, Pole

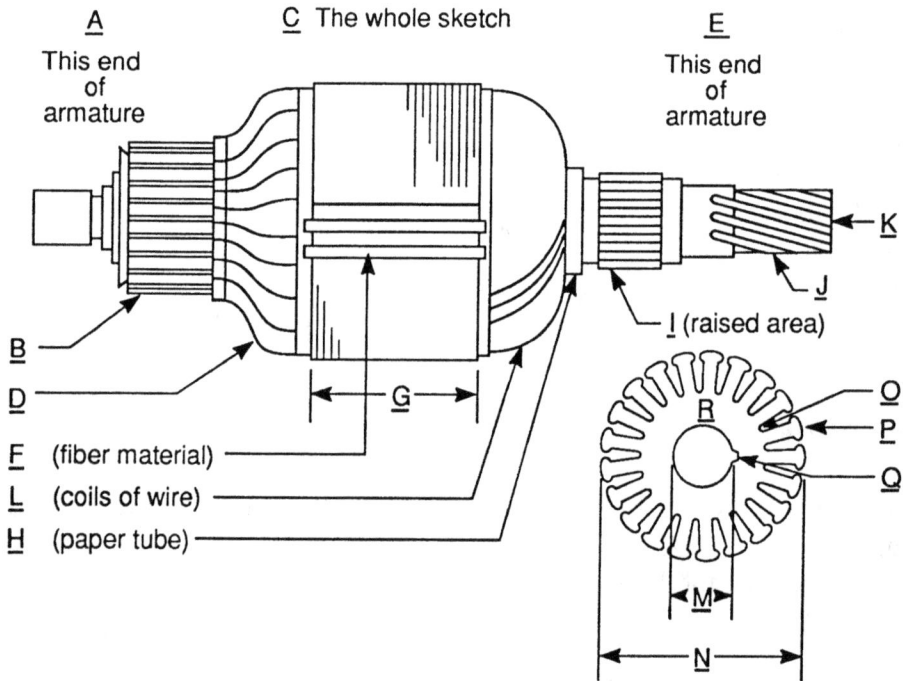

FIGURE 1-2

A	F	K	P
B	G	L	Q
C	H	M	R
D	I	N	
E	J	O	

Answers (not in order):

Coil wire, Comm. End, Pulley End, Spiral Gear, End Turns, Knurl, Lamination, Shaft Insulator, Slot Wedge, Marker Slot, Stack Height, Winding Slot, Commutator, Armature, I.D., Teeth, Shaft, O.D.

EXPERIMENT 2

SEPARATELY-EXCITED DC GENERATOR

-OBJECTIVE-

To properly connect a separately-excited dc generator and to determine its no-load and load characteristics.

-EQUIPMENT-

Dc generator
Dc motor
Variable resistance load
SPST knife switch
3 variable 115 volt dc supplies
2 dc ammeters
Dc voltmeter
VOM
Tachometer

-DISCUSSION-

Separately-excited dc generators are used to produce dc under the control of their field voltage. In doing this the separately-excited dc generator acts as an amplifier, since relatively little power is required to supply the field as compared to the power available from the armature. Most new installations use electronic dc power supply systems instead of separately-excited dc generators, although many existing controlled dc supplies still depend on separately-excited dc generators.

Voltmeter-Ammeter Method

This experiment uses the voltmeter-ammeter method to determine the armature resistance R_A. This method is simply an application of Ohm's law: R_A equals applied low dc voltage divided by produced current. There are two reasons why this method rather than a VOM should be used to measure R_A. First, most R_{AS} are so low that they do not register in a region where a VOM is accurate. Second, the contact resistance of the carbon brush to the commutator surface changes with current magnitude. Resistance measured at near operating current is much more realistic.

Prime Mover

The prime mover, a separately-excited dc motor, has been connected for the student for this experiment. The circuit used is that of Appendix Figure B-1.

V_O Versus I_O Calculations

The output voltage, V_O, versus output current, I_O, characteristics can be calculated using E_G, the internally generated voltage, and the R_A. E_G depends on the speed, field current, and construction of the generator. If these are kept constant, E_G is constant. E_G will equal V_O when there is no I_O, during the no-load condition. The equation to determine the V_O, versus I_O characteristic is:

$$V_O = E_G - I_O \times R_A$$

-PROCEDURE-

1. Record the nameplate data of the motor and generator.
2. Measure the resistance of the armature of the dc generator. Use the voltmeter-ammeter method by applying reduced voltage as shown in Figure 2-1. Vary the voltage to obtain several steps in armature current from 0 to the rated current value shown on the generator's nameplate. The measured voltages and currents will be used to determine the resistance of the armature circuit. Also, measure the resistance with an ohmmeter. Record the readings in Table 2-1.

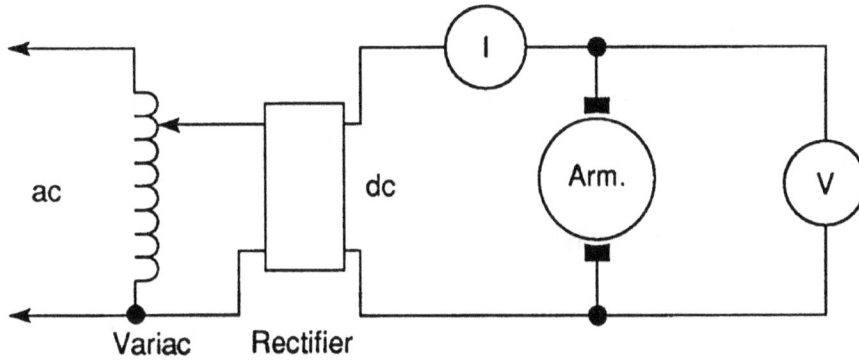

FIGURE 2-1

TABLE 2-1 Armature Resistance

Field Current (amps)	Field Voltage (volts)	V/I (ohms)

Ohmmeter measurement_____ (ohms)

3. Measure series field resistance with the same methods, the voltmeter-ammeter method and the ohmmeter. Record readings in Table 2-2.

TABLE 2-2 Series Field Resistance

Armature Current (amps)	Armature Voltage (volts)	V/I (ohms)

Ohmmeter measurement_____ (ohms)

4. Measure the shunt field resistance with an ohmmeter. Record the reading in Table 2-3.

TABLE 2-3 Shunt Field Resistance

Ohmmeter measurement_____ (ohms)

5. Connect the machine as a separately-excited shunt generator as shown in Figure 2-2, but with the load disconnected.

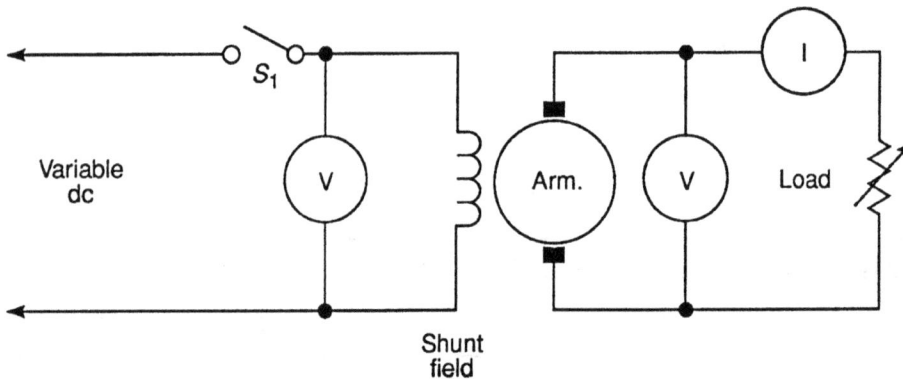

FIGURE 2-2

6. Read Appendix C, page 120, on motor runaway.
7. Have the circuit approved by the instructor.
8. Set the variable dc supply to 0 field voltage and leave S_1 open. Turn on the prime mover and operate it at the generator's rated speed. The generator is now operating without field and the voltage measured at its armature output is due to the residual magnetism in the machine. Record this voltage in Table 2-4.

TABLE 2-4 Generator No-Load Output Versus Field Voltage

	Field Voltage (volts)	Armature Voltage (volts)
Increasing field voltage	0	
Decreasing field voltage		
	0	
Increasing field voltage with field connections reversed		

Speed_____ (rpm)

9. Close S_1 and **slowly** increase the field voltage to about 10 percent of its rated value. Observe the effect of the change of field voltage on output voltage. Now, vary the field voltage from 10 percent to 125 percent of rated in several steps, recording the output voltage for each step in Table 2-4. Be careful to keep the generator rotating at its rated rpm for these steps.

10. Stop the generator from rotating. Open S₁, reverse the field voltage polarity, and reduce the field supply voltage to 0 volts. Rotate the generator at rated rpm. Closing S₁, note the effect on output voltage as field voltage is **slowly** increased from 0 to a small fraction of rated field voltage. Then, increase field voltage in several more steps to 125 percent of rated value, recording output voltage for each step in Table 2-4. Be careful to keep the generator rotating at its rated rpm for these steps.

11. With no load connected to the generator and the generator rotating at rated rpm, set the field voltage so that the generator produces approximately 110 percent of rated output voltage. Apply a load to the generator and vary it so that the generator produces from 0 to 125 percent of the generator's rated output current. Use several steps and record the load voltage and load current for each step in Table 2-5. Do not change the field voltage once it is set and be careful to keep the generator rotating at its rated rpm for these steps.

TABLE 2-5 Loaded Generator Output

Armature Current (amps)	Armature Voltage (volts)

Speed_____ (rpm)

Field Voltage_____ (volts)

-QUESTIONS-

1. Calculate and record the armature and series field resistances. Compare the resistances found by the voltmeter-ammeter method with those found by the ohmmeter.
2. Plot the output voltage as a function of field current for the data of Procedures 9 and 10.
3. What differences occurred as a result of the reversed shunt field polarity in Procedure 10?
4. Plot output voltage as a function of load current using the measured data of Procedure 11. Then predict output voltage for each load current measured, basing your calculations on armature resistance, E_G, and load current. Plot the calculated values on the same graph as the measured values. Compare the calculated and measured values of output voltage. Use the percentage difference equation defined in Appendix A, page 118, to help in the comparison.

EXPERIMENT 3

SHUNT AND COMPOUND
DC GENERATORS

-OBJECTIVE-

To determine the no-load and load characteristics of shunt and compound dc generators.

-EQUIPMENT-

Dc generator
Dc motor
Variable resistance load
SPST knife switch
2 variable 115 volt dc supplies
Dc ammeter
Dc voltmeter
VOM
Tachometer

-DISCUSSION-

Shunt and compound generators are self-exciting. Both produce voltage for their own fields without requiring another dc supply, as the separately-excited dc generator does. Shunt and compound generators have several advantages over the separately-excited generator. First the shunt and compound generators are simpler, since one less dc supply is required. Second, on the compound generator, it is possible to custom fit the output voltage V_O, versus output current, I_O, characteristic curve. Self-exciting generators are used where fixed non-controllable dc supplies are required, as might be the case in supplying a dc power bus.

Compound generators can be connected either cumulatively or differentially. The cumulative compound generator is more common and practical than the differential compound generator. In it, the series winding is connected so that its field adds to the shunt field. The series field produces a greater internally generated armature voltage as the armature current increases. The increased internally generated armature voltage compensates for the armature resistance voltage drop owing to armature current.

The differential compound generator is rarely used, since its V_O drops rapidly as I_O increases.

Shunt Field Rheostat

A shunt field rheostat may be used to control the current to the shunt field of a shunt or compound generator. Increasing the resistance of the shunt field rheostat decreases the amount of current going to the shunt field and thereby decreases the armature voltage. A shunt field rheostat may be used to set the no-load voltage of a shunt or compound generator. R_1 in Figure 3-1 is a shunt field rheostat.

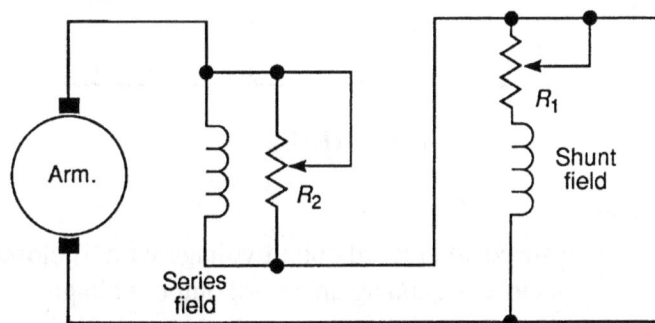

Figure 3-1 Long Shunt Compound Generator

Series Field Diverter

A series field diverter may be used to control the current to the series field of a series or compound generator. Decreasing the resistance to the series field diverter decreases the amount of current going to the series field and thereby decreases the armature-produced voltage. Since the series field only has an effect when the armature is producing current, the series field diverter will only have an effect when the generator is loaded. The series field diverter may be used to set the output voltage of a loaded series or compound generator. R_2 in Figure 3-1 is a series field diverter.

Prime Mover

The prime mover, a separately-excited dc motor, has been connected for the student for this experiment. The circuit used is that of Appendix Figure B-1.

20

-PROCEDURE-

1. Record the nameplate data of the motor and generator.
2. Connect the machine as a shunt generator as shown in Figure 3-2, but with the load disconnected.

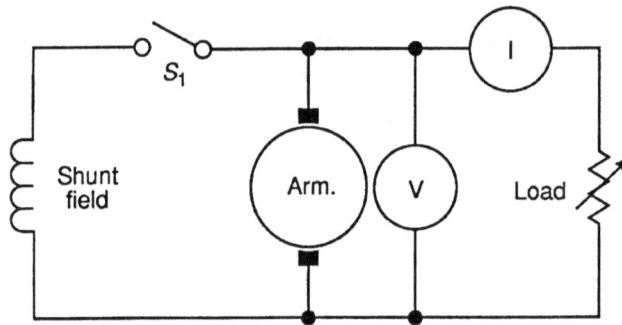

FIGURE 3-2

3. Have the circuit approved by the instructor.
4. Drive the machine at rated speed and record output voltage with S_1 closed. Open S_1, reverse the shunt field connections, close S_1, and again record output voltage.
5. With the shunt field connected to produce proper voltage, connect the generator to the load and vary load current from 0 to 125 percent in several steps recording voltage and load current values.
6. Connect the generator as a short shunt compound generator as shown in Figure 3-3.

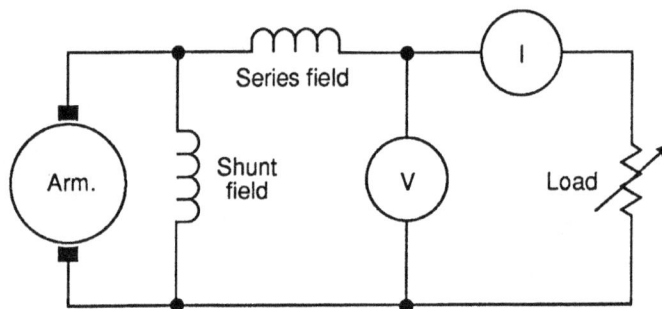

FIGURE 3-3

7. Have the circuit approved by the instructor.

8. Operate the generator at rated speed with no load applied and then apply varying loads from 0 to rated load in several steps. Record output voltage and load current for each step.
9. Reverse the series field connections and repeat the measurements taken in Procedure 8.

-QUESTIONS-

1. Plot output voltage as a function of load current for Procedures 5, 8, and 9.
2. Explain differences in the curves.
3. What differences would be expected between cumulative and differential compound generators?
4. What would be the effect of a shunt field rheostat on the output of a compound generator? What shunt field rheostat resistance would be necessary to halve the current going through the shunt field?
5. What would be the effect of a diverter on the output of a compound generator? What diverter resistance would be necessary to halve the current going through the series field?

EXPERIMENT 4

DC MOTOR CHARACTERISTICS

-OBJECTIVE-

To determine the starting, speed, and torque characteristics of a shunt and a compound dc generator.

-EQUIPMENT-

Dc motor
Starting resistance
SPST knife switch
2 variable 115 volt dc supplies
Dynamometer
Variable resistance load
2 dc ammeters
2 dc voltmeters
VOM
Tachometer

-DISCUSSION-

Many dc motors are used in applications requiring variable speeds. A properly controlled separately-excited dc motor can produce significant torque over a range of a few rpm to thousands of rpm. Shunt and cumulative compound motors can also operate over a wide speed range by using shunt field rheostats and series field diverters along with a controlled dc supply voltage.

Starting Resistance

Large dc motors (and ac motors) often require series resistance at starting to limit starting current. Typically, a starting resistance is chosen that will limit the current to the motor to about 1.5 times the motor's rated full load current. Keeping the current low protects the motor and the electrical system supplying the motor. The value of the starting resistance for a dc shunt motor may be calculated by the following equation:

$$R_S = V_R / [(I_R - V_R/R_F) \times 1.5] - R_A$$

where

R_S = required starting resistance (ohms)
V_R = rated(nameplate) motor input voltage (volts)
I_R = rated(nameplate) motor current (amps)
R_F = shunt field resistance (ohms)
R_A = armature resistance (ohms)

If a resistance of the calculated value were not available, an available one of the next larger size would be used.

Power Equations

Motor input is electrical power. On a dc motor it is simply $V_I \times I_I$ (watts).

where

V_I = input voltage (volts)
I_I = input current (amps)

Motor output is mechanical power. On any motor it may be determined by T x S/5252 (horsepower).

where
T = motor output torque (ft lb)
S = motor speed (rpm)

-PROCEDURE-

1. Record the nameplate data of the motor.
2. Measure the armature resistance of the motor to determine starting resistance requirements.
3. Calculate a value for R_S that will limit starting current to 1.5 times rated.

4. Connect the machine as a shunt motor. Be certain to make solid connections to the shunt field to avoid the possibility of motor runaway. (See Appendix C for a description of motor runaway.) Provide metering for measuring line current supplied to the motor, input voltage, and shunt field current. Include the starting resistance determined in Procedure 3 and a switch to short the starting resistance when the motor is up to speed. See Figure 4-1.

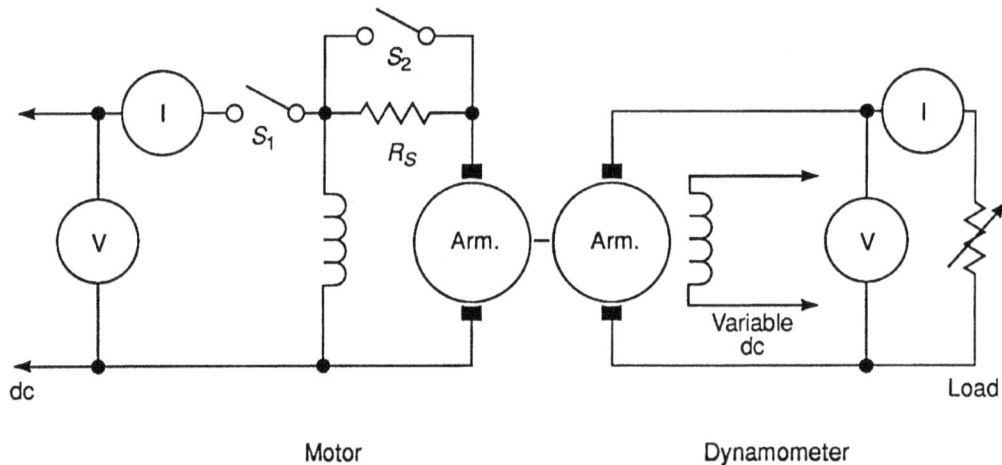

FIGURE 4-1

5. Connect the dynamometer (see Appendix B for dynamometer details) attached to the motor as a separately-excited generator. Connect a voltmeter and ammeter to the dynamometer's armature.
6. Have the circuits approved by the instructor.
7. Disconnect the load from the dynamometer and leave the starting resistor in the circuit for this step. Start the motor and note its direction of rotation (clockwise or counterclockwise as seen from one end of the motor). Stop the motor, reverse the armature connections only, and run the motor again to determine the direction of rotation. Stop the motor, reverse the field leads only, and again determine the direction of rotation.
8. Do a load test on the motor (with its R_S shorted out). Apply load to the dynamometer in five to eight steps until the motor is using rated current. Keep motor input voltage constant during the test. Measure and record input voltage, current, rpm, torque, dynamometer output voltage, and dynamometer output current for each step.
9. Change the motor to a long shunt cumulative compound motor by adding a series field as shown in Figure 4-2.

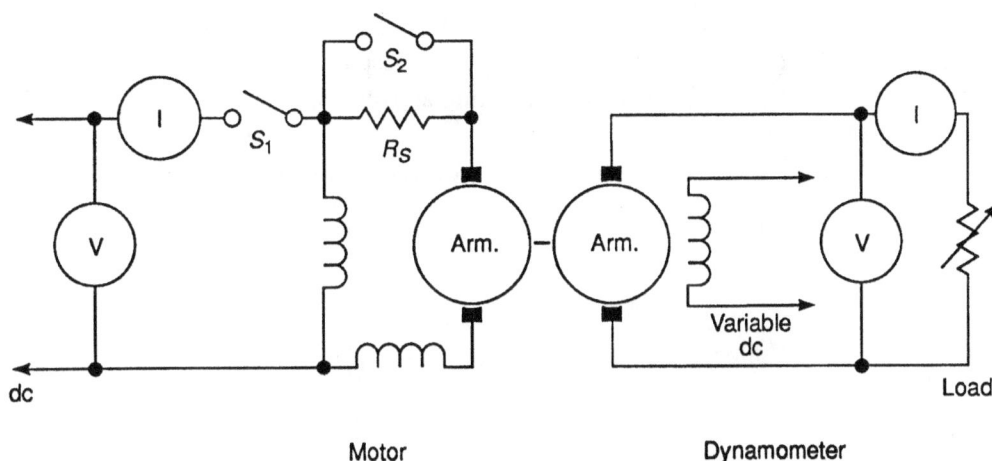

FIGURE 4-2

10. Have the circuit approved by the instructor.
11. Test for the proper series field polarity. First, run the motor with the starting resistor in the circuit and the series field shorted. Note the rotor rotation. Second, run the machine with the starting resistor in the circuit, the shunt field open-circuited, and the series field in the circuit. Note the rotor rotation. If the original compound motor connection were cumulative, the rotor would rotate the same way in each case; if not, the motor was connected differentially and either the series or shunt field connections must be reversed to make cumulative compound connections. Differential compound connections could cause motor runaway, and so should be avoided. (See Appendix C for a description of motor runaway.)
12. Repeat Procedure 8 with the compound motor.

-QUESTIONS-

1. Explain the effects of changing connections in Procedure 7 on the direction of rotation.
2. Plot torque versus armature current for the shunt and compound motors. Use one graph for both curves.
3. Plot the values for speed (rpm) for each motor as a function of armature current. Use one graph for both curves.
4. Plot values of torque of each motor as a function of speed. Use one graph for both curves.
5. Give reasons for similarities and differences between the corresponding curves of Questions 2, 3, and 4.

6. Calculate the output powers of each motor using torque and rpm for each step in Procedures 8 and 12. Then, calculate the dynamometer's output power for each step by multiplying its armature voltage times armature current. Compare the power values and explain differences between them.

7. Compare mechanical powers measured at rated current with nameplate data.

EXPERIMENT 5

DC MOTOR LOSSES, EFFICIENCY, AND BRAKING

-OBJECTIVE-

To determine a dc shunt motor's losses and efficiencies at various loads and to demonstrate methods of motor braking.

-EQUIPMENT-

Dc motor (same as used in Experiment 4)
Braking resistances
DPDT switch
Variable 115 volt dc supply
VOM
Stopwatch

-DISCUSSION-

Losses and Efficiency

A dc motor has fixed and variable losses. Fixed losses include mechanical friction losses, shunt field resistance losses, and magnetic eddy current and hysteresis losses. For these to be "fixed" losses, a reasonable approximation is made that the motor is operating at a constant rated input voltage and at a constant rated speed. Variable losses are the resistance losses to the armature and series field. Since current magnitudes vary as a motor is loaded down, these will vary.

The electrical input to a dc motor is $V_I \times I_I$ (watts).

where

V_I = motor input voltage (volts)
I_I = motor input current (amps)

Based on knowledge of the input electrical voltage and current, the fixed losses, and the armature resistance, the output mechanical power of a shunt motor may be estimated. The fixed losses to a dc motor are very close to the total power consumed when the motor is running at no-load. The following equation will calculate a shunt motor's mechanical output in watts:

$$P_O = V_I \times I_I - P_F - (I_I - V_I/R_F)^2 \times R_A$$

where

P_O = motor output current power (watts)
P_F = motor fixed losses (watts)
R_F = shunt field resistance (ohms)
R_A = armature resistance (ohms)

Percent motor efficiency may then be calculated by the equation:

$$\% \text{ Efficiency} = (P_O/P_I) \times 100\%$$

where

$$P_I = V_I \times I_I = \text{motor input power (watts)}$$

Braking

Both mechanical and electrical motor braking methods may be used. This experiment deals with the electrical methods of plugging and dynamic braking on dc shunt motors.

Plugging stops a motor by temporarily applying reverse voltage to it. To do so on a dc shunt motor, the armature connections are temporarily reversed. This reverses torque on the armature and will stop the motor. Care must be taken that the reversed armature connections are disconnected as soon rotation stops; otherwise, the motor will start rotating in the opposite direction. It is necessary to have a series resistance in the plugging circuit to limit armature current. Typically, this current is limited to about 1.5 times the motor's rated current. The following equation will calculate the plugging resistance for a dc motor:

$$R_P = (2 \times V_R)/(1.5 \times I_R) - R_A$$

where

R_P = plugging braking resistance (ohms)
V_R = rated motor voltage (volts)
I_R = rated motor current (amps)

Dynamic braking stops a dc motor by turning it into a separately-excited dc generator. In operation, the armature is disconnected from the dc supply and reconnected to a resistance, while the shunt field is left connected to the dc supply. As long as the motor turns, it will produce armature voltage and send electrical power to the resistance connected to its armature. Dynamic braking has the advantage that the critical step of removing the braking circuit at the right moment is not required. A resistance of proper size must be chosen to produce good dynamic braking. Typically, a resistance is chosen to limit armature current to about 1.5 times the motor's rated current. The following equation will calculate the resistance required to apply dynamic braking to a dc motor:

$$R_D = V_R/(1.5 \times I_R) - R_A$$

where

R_D = dynamic braking resistance (ohms)

-PROCEDURE-

1. Recopy the load data of the shunt motor tested in Experiment 4.
2. Calculate the plugging resistance and dynamic resistance to limit current to 1.5 times rated. If the exact values of R_P and R_D are not available, choose the next larger sizes available.
3. Construct the plugging circuit of Figure 5-1. Use a DPDT knife switch as a reversing switch and include the plugging resistance value calculated in Procedure 2.

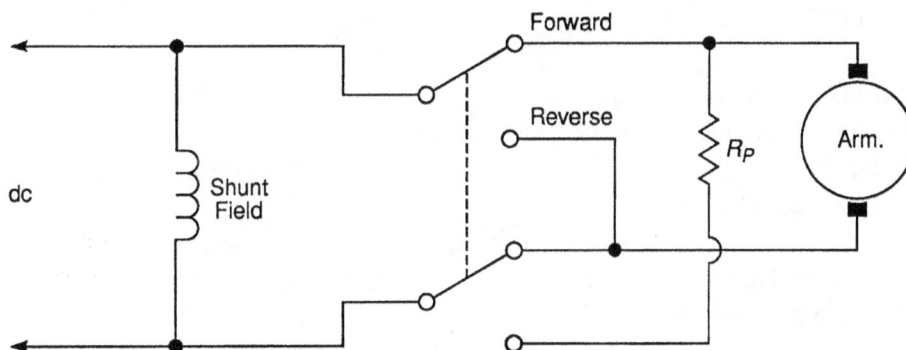

FIGURE 5-1

4. Have the circuit approved by the instructor.
5. Operate the motor with rated voltage and no load; then open the knife switch and let the

motor coast to a stop. Note the time required for the motor to stop. Do several trials.

6. Again operate the motor with rated voltage and no load. However, on this trial move the knife switch rapidly from the normal position to the reverse position. Note the time required for the motor to stop. Do several trials.

7. Reconnect the knife switch to provide dynamic braking as shown in Figure 5-2.

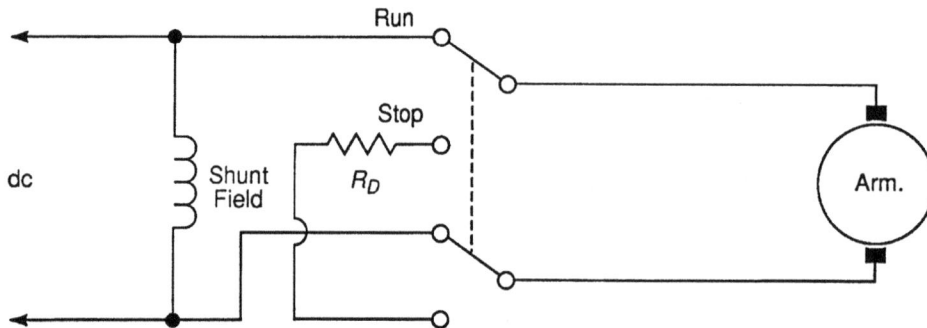

FIGURE 5-2

8. Have the circuit approved by the instructor.

9. Again operate the motor with rated voltage and no load. Then, rapidly move the switch from the normal running position to the dynamic braking position. Note the time required for the motor to stop. Do several trials.

-QUESTIONS-

1. Use the data of Experiment 4 to:
   a. Determine the fixed losses of the motor at rated speed.
   b. Calculate the armature copper losses of the motor for currents of 20, 40, 60, 80, 100, 120, and 140 percent of rated current.
   c. Using the fixed losses determined in Question 1a and the armature copper losses determined in 1b, calculate the percentage efficiency and power output of the motor at each current of Question 1b.
   d. Using torque, speed, voltage in, and current in data, again determine the motor's percentage efficiency for each power output of the motor.
   e. On the same piece of graph paper, plot two curves of percentage efficiency versus mechanical power output. Use the results of Question 1c for one curve and the results of Question 1d for the other. Compare the curves.

2. Compare the motor's coasting, plugging, and dynamic braking stopping times. Comment concerning the usefulness of each method and list the precautions that might be necessary with each method.

PART II

AC THREE-PHASE POWER AND

TRANSFORMER EXPERIMENTS

EXPERIMENT 6

THREE-PHASE POWER MEASUREMENT

-OBJECTIVE-

To demonstrate methods of measuring three-phase power.

-EQUIPMENT-

60 watt light bulb
3 100 watt light bulbs
3 light bulb sockets
2 15 μF, 115 volt ac capacitors
Fixed 115 volt three-phase supply
Ac clamp-type ammeter
VOM
3 identical wattmeters

-DISCUSSION-

Three-phase power may be measured with one, two, or three wattmeters.

Three-Wattmeter Methods

Three-wattmeter methods are accurate with balanced and unbalanced three-phase sources and loads. Loads may be of any power factor. Total power is the sum of the three wattmeter readings.

Figures 6-1 and 6-2 show three-wattmeter methods where the wattmeters are directly connected across the load of each phase.

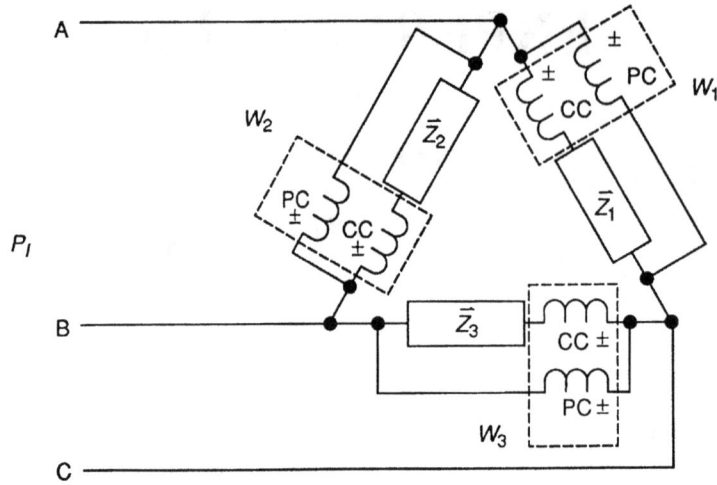

FIGURE 6-1 Directly Connected Three-Wattmeter Method to Measure Power to a Delta Load

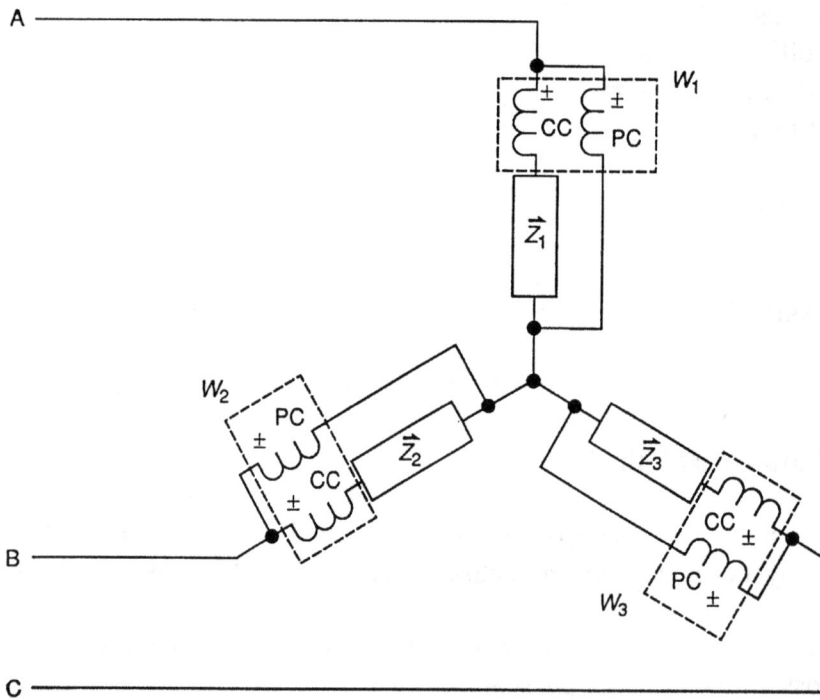

FIGURE 6-2 Directly Connected Three-Wattmeter Method to Measure Power to a Wye Load

It is often impossible or inconvenient to connect wattmeters directly to each phase load. In such cases the remotely connected three-wattmeter method may be used. This method requires that all wattmeters have the same potential coil resistance. Ideally, identical wattmeters should be used for all three phases. The method is shown in Figure 6-3.

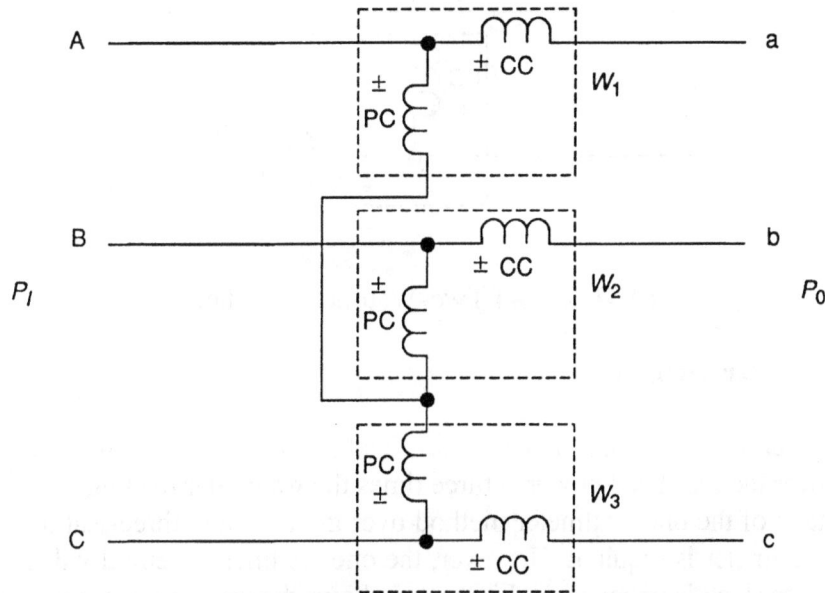

FIGURE 6-3 Remotely Connected Three-Wattmeter Method

Two-Wattmeter Method

The two-wattmeter method is also accurate with balanced and unbalanced three-phase sources and loads. Again, loads may be of any power factor. Total power is the sum or difference of the two-wattmeter readings, depending on the load power factor. If the power factor of the load is greater than 0.5, the wattmeter readings should be added to give the total power to the load. If the power factor of the load is less than 0.5, the smaller wattmeter reading should be subtracted from the larger to give the total power to the load.

The advantage of the two-wattmeter over the three-wattmeter method is that one less wattmeter is required. Since wattmeters are expensive, this can be important. However, the two-wattmeter method can be confusing, especially when the load is unbalanced and/or has a changing power factor. Figure 6-4 shows the two-wattmeter method.

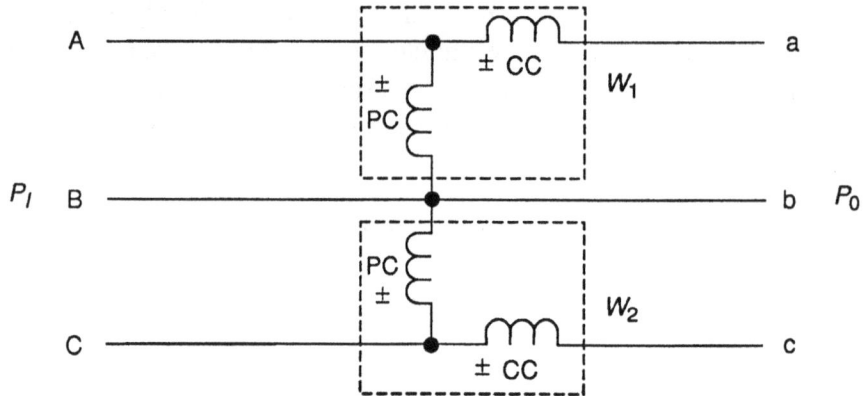

FIGURE 6-4 Two-Wattmeter Method

One-Wattmeter Method

The one-wattmeter method is also called the artificial neutral method. It will measure power to loads of all power factors. Total power is three times the wattmeter reading.

The advantage of the one-wattmeter method over the two- and three-wattmeter methods is that only one wattmeter is required. However, the one-wattmeter method will be inaccurate if either the source or load is unbalanced. Figure 6-5 shows the one-wattmeter method.

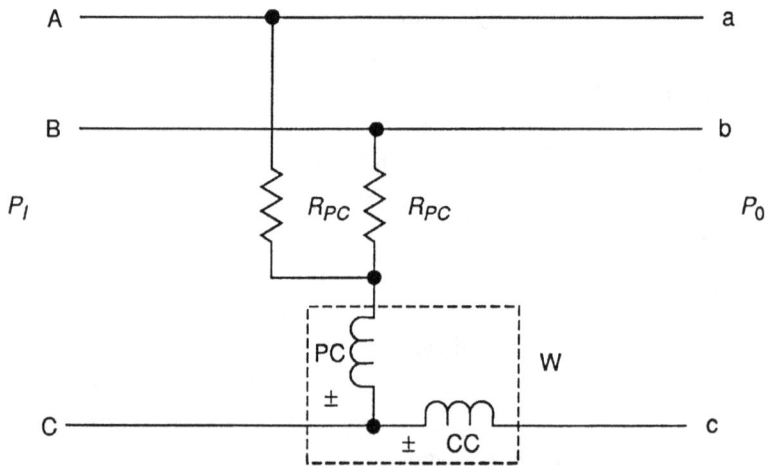

FIGURE 6-5 One-Wattmeter Method

Wattmeter Polarity

Unlike most ac meters, wattmeters have polarity marks on their terminals. This is because power has polarity. If a wattmeter is connected with the wrong polarity voltage coil relative to current coil, it will try to indicate a negative watt value. Most analog wattmeters can't do that. Wattmeter potential and current coils are each marked on one end with a $\pm$. Those marks can also be seen in the diagrams of figures 6-1 to 6-5. The experimenter should mark the polarity of wattmeters in all circuit diagrams. Circuits with wattmeter polarity marks are easier to understand and build.

-PROCEDURE-

1. Connect the circuit of the three-wattmeter method of Figure 6-3 to the balanced, purely resistive load circuit of Figure 6-6. Use the 100 watt bulbs as the load resistances. Connect source A to load a, source B to load b and source C to load c. Put paper labels on the A, B, C, a, b, c terminals and on the wattmeters. The labels will make circuit construction and operation easier. Have the circuit approved by the instructor before initially energizing it.

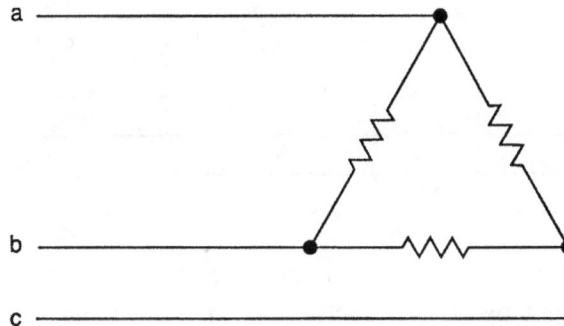

FIGURE 6-6

2. If using a clamp-type ammeter, it will probably be necessary to put extra conductor loops through the clamp for some readings. This procedure is explained in Appendix B, page 115.
3. Energize the circuit. Measure all line currents, phase currents, line voltages, and wattmeter readings. Record these in Tables 6-1 and 6-2.

38

TABLE 6-1 Voltage, Current, and Calculated Power

	Resistive Balanced Load	Resistive Unbalanced Load	Leading pf Unbalanced Load
I_A (amps)			
I_B "			
I_C "			
I_{ac} "			
I_{cb} "			
I_{ba} "			
V_{ac} (volts)			
V_{cb} "			
V_{ba} "			
Calculated Total Power (watts)			

TABLE 6-2 Balanced, Purely Resistive Load

	Three-Wattmeter Method	Two-Wattmeter Method
W_1 (watts)		
W_2 "		
W_3 "		xxxxxxxxxxxxxxxxxxxxxxxxx
Total Power "		

4. Use the data presented in Table 6-1 to calculate the power to each bulb by multiplying respective bulb voltages by bulb currents. Sum the bulb powers to get the total power and record that number as the "calculated total power" in Table 6-1.
5. Add the three wattmeter readings and record that number as "total power" in Table 6-2. The "calculated total power" of Table 6-1 and the "total power" of Table 6-2 should be close. If they aren't, double check the readings.
6. Replace the ba phase 100 watt bulb with the 60 watt bulb. Now the load is unbalanced, although it is still purely resistive. Repeat Procedures 1 to 5, recording the results in Tables 6-1 and 6-3.

TABLE 6-3 Unbalanced, Purely Resistive Load

	Three-Wattmeter Method	Two-Wattmeter Method
W$_1$ (watts)		
W$_2$ "		
W$_3$ "		xxxxxxxxxxxxxxxxxxxxxxx
Total Power "		

7. Replace the cb phase 100 watt bulb and the ac phase 100 watt bulb with 15 μF capacitors. Now the load is unbalanced and has a leading power factor of less than 0.5. Repeat Procedures 1 to 5, recording the results in Tables 6-1 and 6-4.

TABLE 6-4 Unbalanced, Leading Power Factor Load

	Three-Wattmeter Method	Two-Wattmeter Method
W$_1$ (watts)		
W$_2$ "		
W$_3$ "		xxxxxxxxxxxxxxxxxxxxxxx
Total Power "		

8. Repeat Procedures 1 to 7 with the two-wattmeter method of Figure 6-4. As before, connect source A to load a, source B to load b, and source C to load c. Have the circuit approved by the instructor before initially energizing it. It won't be necessary to take voltage and current readings again, because they shouldn't have changed. It should be necessary to reverse connections to one wattmeter's current or voltage coils to make that wattmeter read on the positive scale when testing the leading power factor circuit. Record the wattmeter readings in Tables 6-2, 6-3, and 6-4.

-QUESTIONS-

1. Why should identical wattmeters be used in the two- and three-wattmeter methods?
2. It is possible to use the data of the three-wattmeter method to show the one-wattmeter method. For balanced loads the total power should equal three times each wattmeter reading. Show how this works with the balanced load three-wattmeter data and how it doesn't work with the unbalanced load data.

40

3. Neatly draw and label one circuit diagram that shows power being measured to a resistive three-phase wye load simultaneously by the one-wattmeter method, the two-wattmeter method, and a three-wattmeter method. The circuit should contain six wattmeters. Label all current coils as CC, potential coils as PC and coil polarities as $\pm$.

EXPERIMENT 7

TRANSFORMER TESTS
AND CHARACTERISTICS

-OBJECTIVE-

To determine a transformer's equivalent circuit and to use this circuit to determine its efficiency and voltage regulation.

-EQUIPMENT-

Filament transformer
Variable 115 volt ac supply
Ac ammeter
Ac voltmeter
VOM
Wattmeter

-DISCUSSION-

Marking Transformer Leads

Often transformer leads are not clearly marked as to their windings and winding polarity.

To distinguish one winding from another, use an ohmmeter to separate windings by continuity or the lack of it. Then, mark the leads of each winding with the same letter and a different number (i.e., X1, X2, X3, etc.). Finally, make a sketch of the transformer with its windings.

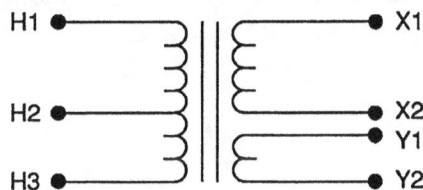

FIGURE 7-1 Example

To determine winding polarity, connect a low ac voltage from a variac to two leads of one winding of the transformer. Measure the voltages produced across the terminals of the different windings and record these voltages on the sketch of the transformer.

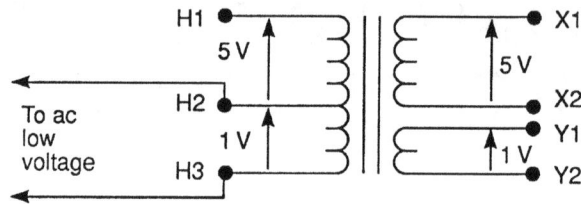

FIGURE 7-2 Example

Put a dot on one lead of one winding in the sketch. Using a voltmeter and a jumper wire, determine if the resultant voltage of that winding in series with another is the sum or difference of their voltages. Put a dot on the second winding so that its voltage adds with that of the first winding, when connected dotted lead to non-dotted lead. Determine dots for each winding, relative to the original dotted winding.

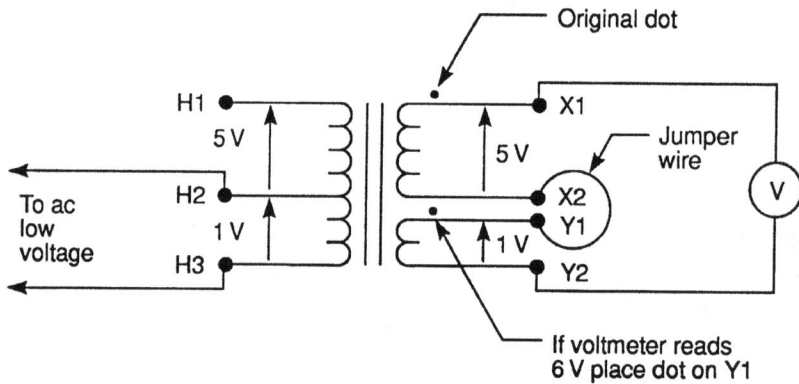

FIGURE 7-3 Example

Renumber and reletter the leads. The highest voltage winding uses H; others use X, Y, Z, and so on. The dotted leads usually have odd numbers.

43

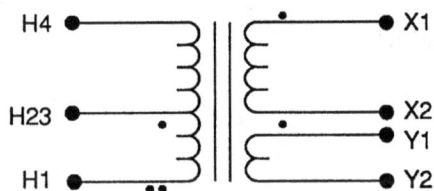

FIGURE 7-4 Example

Transformer Equivalent Circuit

A transformer can be represented as a resistance and inductance in series with an ideal transformer.

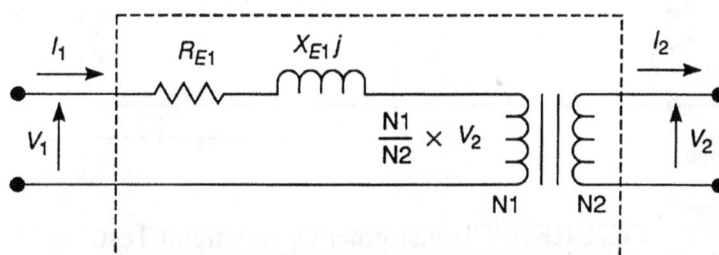

FIGURE 7-5 Equivalent Circuit of a Transformer

Short-Circuit Test

A short-circuit test determines the R_{E1} and X_{E1} of a transformer and the transformer's full load copper losses.

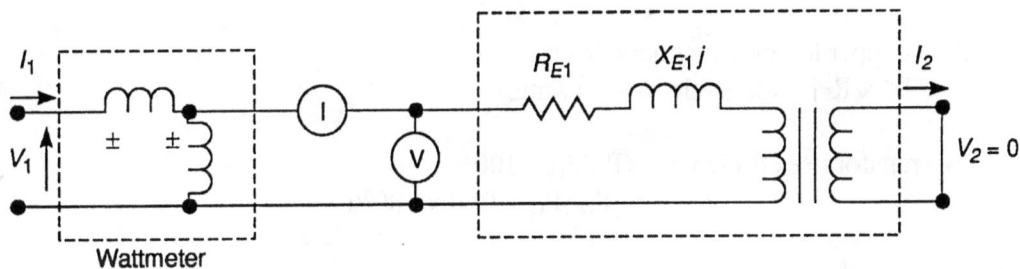

FIGURE 7-6 Transformer Short-Circuit Test

Apply a low voltage to the input of the transformer, so as to supply rated current to the transformer. The wattmeter reads the copper losses of the transformer at rated current.

$$R_{E1} = (\text{Wattmeter reading})/(I_1^2)$$

$$Z_{E1} = V_1/I_1$$

$$X_{E1} = (Z_{E1}^2 - R_{E1}^2)^{.5}$$

Open-Circuit Test

An open-circuit test determines the magnetic (core) losses of a transformer at rated voltage.

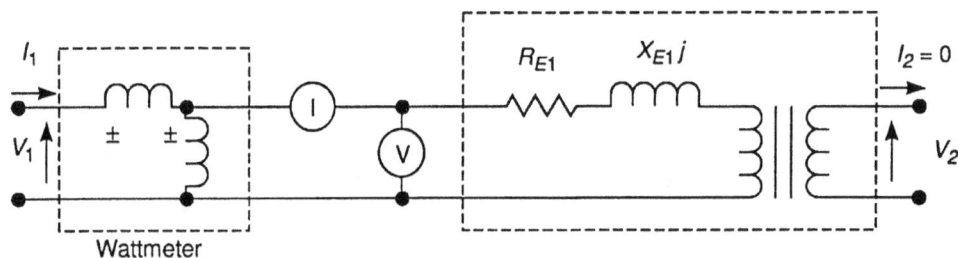

FIGURE 7-7 Transformer Open-Circuit Test

Apply rated voltage to the input of the transformer. The wattmeter reads magnetic (core) losses. The ammeter reads the magnetizing current.

Transformer Efficiency

$$P_O = \text{transformer output power} = V_O \text{ x } I_O \text{ x } \cos\Theta \text{ (watts)}$$

$$P_I = \text{transformer input power} = P_O + P_L \text{ (watts)}$$

$$P_L = \text{copper losses} + \text{magnetic losses}$$
$$= I_1^2 \text{ x } R_{E1} + \text{magnetic losses (watts)}$$

$$\% \text{ Transformer efficiency} = (P_O/P_I) \text{ x } 100\%$$
$$= [P_O/(P_O + P_L)] \text{ x } 100\%$$

Voltage Regulation

$$\% \text{ Voltage Regulation} = [(V_{NL} - V_{FL})/V_{FL}] \text{ x } 100\%$$

-PROCEDURE-

1. Copy or estimate (from similar transformers or catalogs) the nameplate rating of the transformer.
2. Draw the transformer's schematic. Label the schematic and transformer leads with H, X, and Y as appropriate.
3. Draw a schematic for each of the following tests (include meters).
    a. Short-circuit test. Refer all impedance to the high-voltage side of the transformer. If the transformer has more than two windings, consider the transformer's low-voltage output to be the low-voltage windings connected constructively in series.
    b. Open-circuit test.
4. Do the tests.

-QUESTIONS-

1. Calculate the transformer's R_{E1} and X_{E1}.
2. What are the transformer's full load copper losses and core losses?
3. Using the copper and core losses, calculate the transformer's full load efficiency.
4. Using the X_{E1} and R_{E1}, calculate the transformer's full load voltage regulation.

EXPERIMENT 8

TRANSFORMERS AND AUTOTRANSFORMERS

-OBJECTIVE-

To compare the operation of a transformer with isolated windings to its operation when connected as an autotransformer.

-EQUIPMENT-

Filament transformer (the same as in Experiment 7)
Variable resistance load
Variable 115 volt ac supply
Ac ammeter
Ac voltmeter
VOM
Wattmeter

-DISCUSSION-

A transformer can be connected as an autotransformer by connecting its secondary winding(s) in series with its primary winding(s).

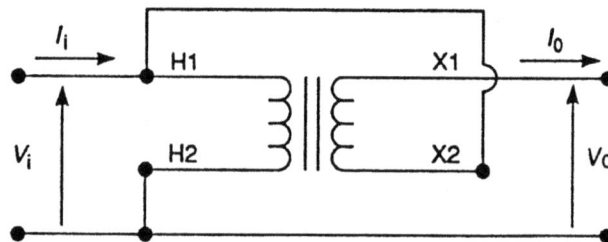

FIGURE 8-1 Autotransformer Connection

A transformer connected as an autotransformer has a greater output rating (V_O x I_O) because the transformer only carries a part of the output power. For the same reason, the voltage regulation and efficiency improve when a transformer is connected as an autotransformer.

-PROCEDURE-

1. Draw a circuit to load test the transformer when it is connected with isolated windings. Connect the load to the low-voltage side of the transformer. Rated voltage should be applied to the high-voltage side. Include meters in your circuit.
2. Construct the circuit and test the transformer over a range of outputs from 0 to rated output current.
3. Draw a circuit to load test the transformer when the transformer is connected as a step-up autotransformer. Again, the circuit should operate with rated voltage applied to the high-voltage side of the transformer.
4. Construct the circuit and test the autotransformer over a range of outputs from 0 to rated output.

-QUESTIONS-

1. Using the R_{E1}, X_{E1}, and the magnetic losses found in Experiment 7, calculate the low-voltage output voltages and the percentage efficiencies of the transformer in Procedure 1. Compare the calculated results with those of Procedure 2 on a graph of output voltage and percentage efficiency versus output power.
2. Using the R_{E1}, X_{E1}, and the magnetic losses found in Experiment 7, calculate output voltages and the percentage efficiencies of the autotransformer of Procedure 3. Compare the calculated results with those of Procedure 4 on a graph of output voltage and percentage efficiency versus output power (watts). (Hint: To make the calculations easier, approximate the magnitude of Z_{E1} as a pure resistance when calculating voltages and currents.)

EXPERIMENT 9

THREE-PHASE TRANSFORMER CONNECTIONS

-OBJECTIVE-

To demonstrate various ways of connecting single-phase transformers to transform three-phase.

-EQUIPMENT-

3 identical single-phase transformers
1 small amperage fuse
Three-phase Wye resistance load (leg resistances approximately twice each transformer's rated V_O/I_O)
Fixed 115 volt three-phase supply
Ac clamp-type ammeter
VOM

-DISCUSSION-

Three-Transformer Circuits for Transforming Three-Phase

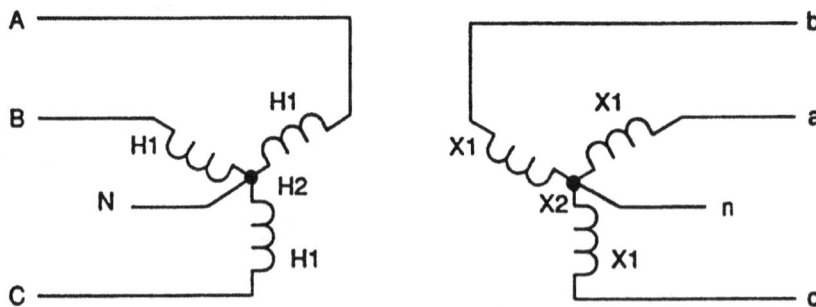

FIGURE 9-1 Wye-Wye

49

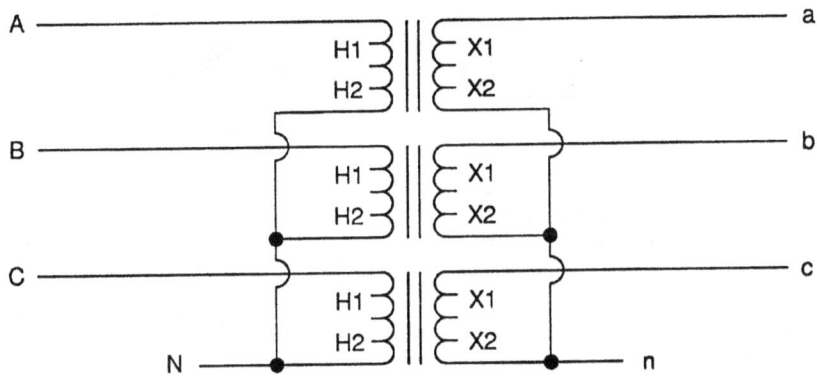

FIGURE 9-2 Other Way of Drawing a Wye-Wye

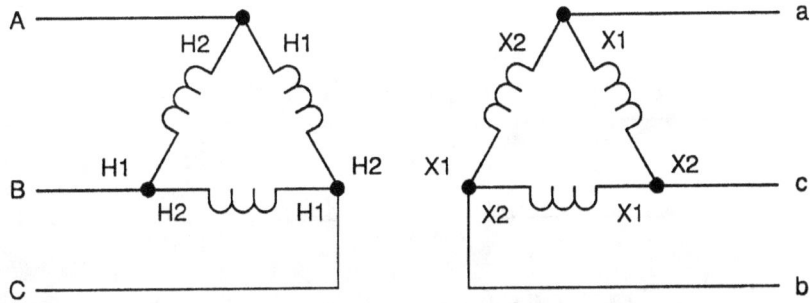

FIGURE 9-3 Delta-Delta

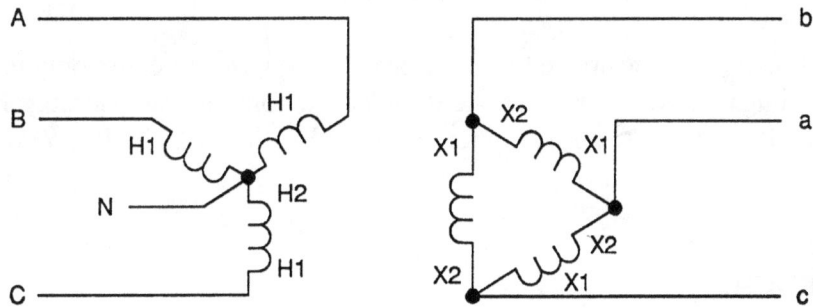

FIGURE 9-4 Wye-Delta

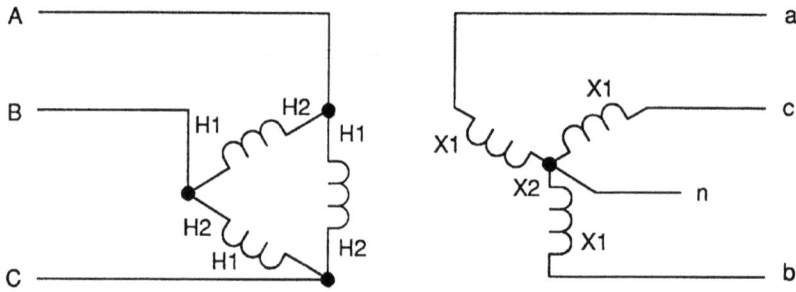

FIGURE 9-5 Delta-Wye

Two-Transformer Circuits for Transforming Three-Phase

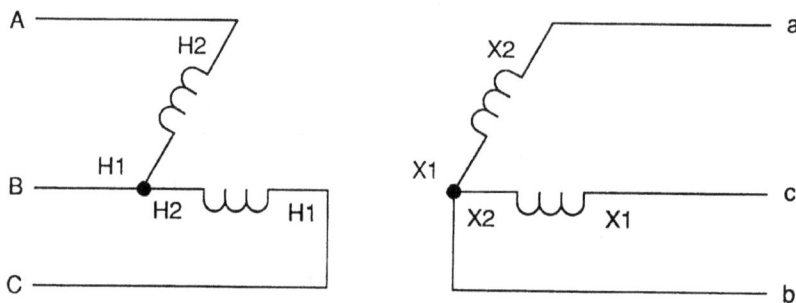

FIGURE 9-6 Open Delta or V-V

One Transformer Circuits for Transforming Three-Phase

Transformers specifically constructed to transform three-phase are commonly used. The input and output windings are wound around one multilegged core, but the windings have the same connection circuits as the three-transformer circuits: Wye-Wye, Delta-Delta, Wye-Delta, and Delta-Wye.

-PROCEDURE-

1. Determine the winding polarities on all transformers. Mark each transformer with an appropriate H1, H2, X1, and X2.
2. Following the circuit diagram of the discussion section, connect three transformers in Wye-Wye to the Wye load. Connect the transformers' input neutral to the source's neutral and the transformers' output neutral to the load's neutral. Pay careful attention to the windings' polarities.

a. Apply voltage. Measure all transformer input and output voltages and currents (line, phase, and neutral).

b. Reverse the output connections of one transformer and repeat Procedure 2a.

c. With the output connections of one transformer still reversed, remove the neutral wire to the load and repeat Procedure 2a. Also measure the voltage from the transformers' output neutral to the load neutral.

3. Following the circuit diagram of the discussion section, connect the three transformer inputs in Delta. Then connect the transformer outputs in open Delta with the load disconnected. This is shown in Figure 9-7. Pay careful attention to the windings' polarities.

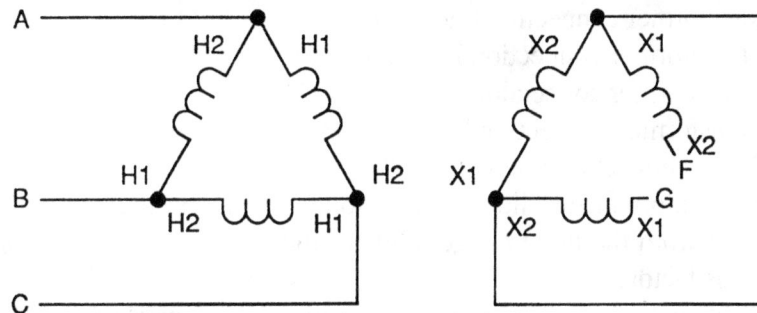

FIGURE 9-7

a. Connect a voltmeter from F to G.

b. Apply voltage. Measure the voltage between F and G. If the windings are properly connected, the measured voltage will be small. Theoretically, the voltage would be 0, but a slight voltage is generated by transformer harmonics.

c. Reverse the output connections of one transformer. Apply voltage. Measure the voltage between F and G. If the windings are properly connected for this step, the measured voltage should be twice that of a transformer output voltage. If the points F and G were now connected, large undesirable currents would circulate through the transformers, tripping a circuit breaker or destroying the transformers.

d. Return the reversed transformer output to its original connections. Once again, the measured voltage across F and G should be small. Provided that the voltage is small, connect F and G through the fuse. If the connections are correct, the fuse will not blow and the Delta can be closed.

e. Connect the transformers' output to the three-phase Wye load. Apply voltage. Measure all transformer input and output voltages and currents.

4. Following the circuit diagram of the discussion section, connect the transformers in open Delta to the load. Measure the input and output line voltages.

52

5. Following the circuit diagram of the discussion section, connect the transformers Delta-Wye to the Wye load with the transformer neutral connected to the load neutral. Measure the input and output line voltages.

-QUESTIONS-

1. What was the ratio of the individual transformers' input voltages to output voltages?
2. What was the ratio of the
   a. Wye-Wye transformer connection line voltages?
   b. Wye-Wye transformer connection line currents?
   c. Delta-Delta transformer connection line voltages?
   d. Delta-Delta transformer connection line currents?
   e. Delta-Wye transformer connection line voltages?
   f. V-V transformer connection line voltages?
3. By what multiplication factor did the line voltage ratio of the Delta-Wye transformer connections differ from the line voltage ratios of the other transformer connections? What is the cause of this factor?
4. Write detailed instructions for connecting three identical unmarked transformers in Wye-Delta. Assume the reader of your instructions knows how to read circuit diagrams and use a VOM before he or she reads your instructions.

PART III

AC MOTOR AND ALTERNATOR

EXPERIMENTS

EXPERIMENT 10

THREE-PHASE SQUIRREL CAGE INDUCTION MOTOR CHARACTERISTICS

-OBJECTIVE-

To determine the curves of torque versus speed, power factor versus mechanical load, and efficiency versus mechanical load of a three-phase squirrel cage induction motor by load testing.

-EQUIPMENT-

Three-phase squirrel cage induction motor
Fixed 115 volt three-phase ac supply
Variable 115 volt dc supply
Dynamometer
Variable resistance load
Ac ammeter
Ac voltmeter
Dc ammeter
Dc voltmeter
VOM
Wattmeter
2 resistances equal to the wattmeter's potential coil resistance
Tachometer
SPST switch

-DISCUSSION-

Torque Versus Speed Curves

When sizing a motor for use, a major concern is, "Will the motor stall or be overloaded?" An equipment designer who works with torque versus speed curves and has knowledge of the mechanical load can choose a motor that will not stall or be overloaded.

Often, curves of a particular motor are not available. The equipment designer should use the motor's National Electrical Manufacturers Association (NEMA) Design Classification with standard curves. Some standard curves are shown in Figure 10-1.

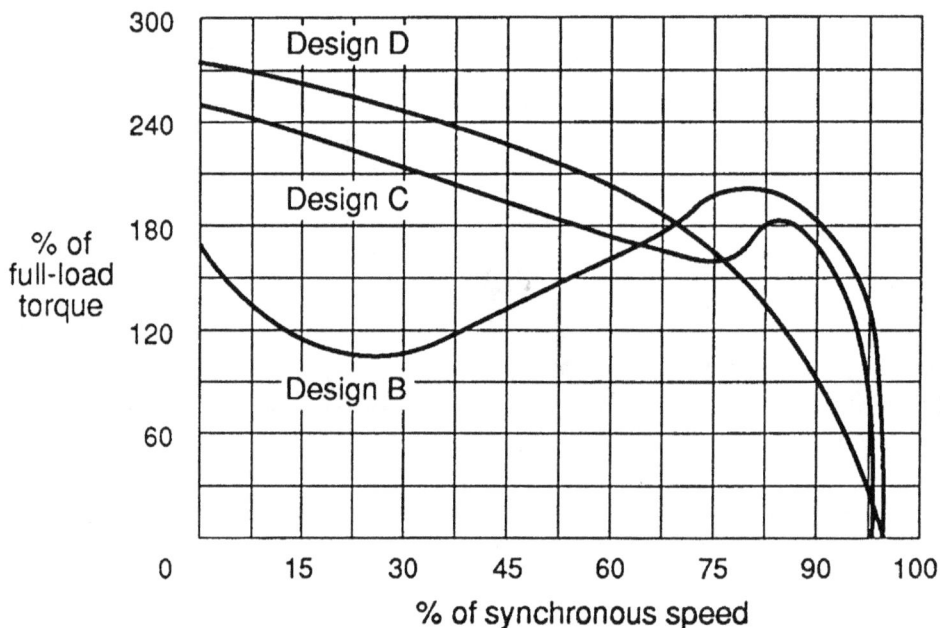

FIGURE 10-1 Standard Motor Curves

Power Factor and Efficiency Versus Load Curves

After sizing a motor so that it will not stall, a designer should consider the operating costs of the motor.

A motor that is too large will waste energy. Specifically, an oversized motor will have greater rotation and magnetic losses than a properly sized motor. It will use more vars and so will increase the reactive power drawn from the power system. It will also require larger and therefore more expensive power leads, switchgear, and fuses or circuit breakers.

Approximate curves of power factor and efficiency versus load are given in Figure 10-2.

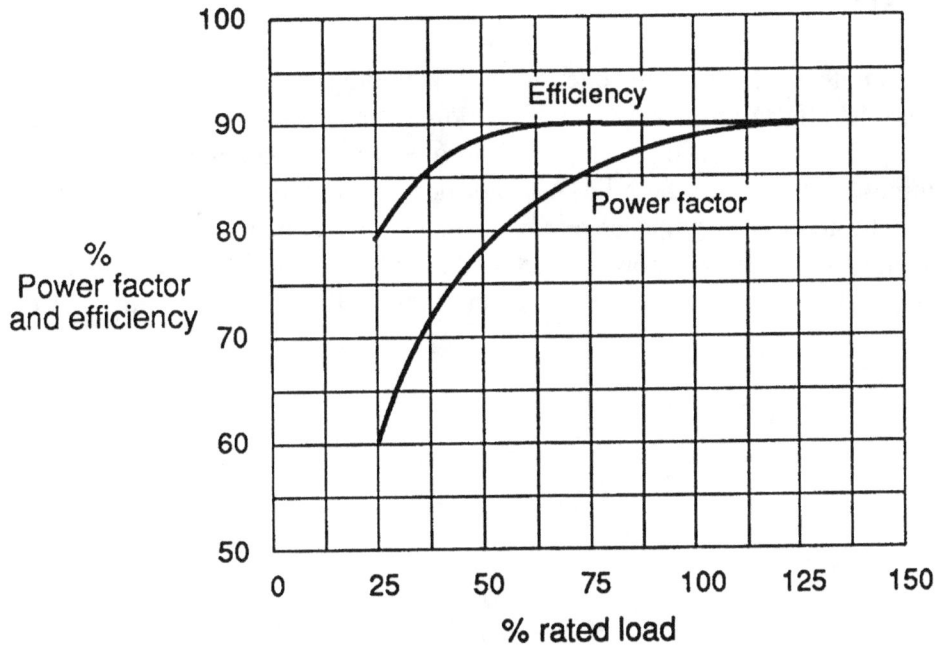

FIGURE 10-2 Typical Motor Curves

-PROCEDURE-

1. Copy the nameplate data of the three-phase squirrel cage induction motor.
2. Draw a circuit diagram of the motor powered by a three-phase source. Connect the motor's diagram by a dotted line (indicating the shaft) to a circuit diagram of the dynamometer. Include meters in your diagram. There should be meters to measure ac voltage, current, and power to the three-phase induction motor. Use the one-wattmeter method of Experiment 6 to measure power. Put a shorting switch across the wattmeter's current coil and the ac ammeter to protect them during motor starting. There should be meters to measure dc voltage and current from the armature of the dynamometer.
3. Construct the circuit and have it approved by the instructor.
4. Short the current coil of the wattmeter and the ac ammeter.
5. Energize the motor and note its rotation.
6. Stop the motor and reverse two of the three power input leads.
7. Energize the motor again and note its rotation.
8. Open the short on the current coil of the wattmeter and ac ammeter. Do a load test on the motor. Record each meter reading as the motor is loaded in 10 steps from no-load to 115 percent of rated load. Consider the motor to be at rated load when it is drawing rated current.

58

-QUESTIONS-

1. Draw a complete diagram of the circuits used, including meters.
2. What was the effect of reversing two of the power input leads? Why did this happen?
3. Draw a graph of percentage full load torque versus percentage synchronous speed. Superimpose the curve of the NEMA motor design that most closely approximates it onto the graph. From the superimposed curve, estimate the starting torque of the tested motor.
4. Draw a graph of percentage power factor and percentage efficiency versus percentage rated load. Superimpose the curves of the discussion section onto the graphs. Compare the curves.

EXPERIMENT 11

INDUCTION GENERATOR

-OBJECTIVE-

To demonstrate the operation of a three-phase induction motor as an electrical power source.

-EQUIPMENT-

Three-phase squirrel cage induction motor (the same as in Experiment 10)
Fixed 115 volt three-phase ac supply
2 variable 115 volt dc supplies
Dynamometer
Ac ammeter
Ac voltmeter
Dc ammeter
Dc voltmeter
VOM
Wattmeter
2 resistances equal to the wattmeter's potential coil resistance
Tachometer
SPST switch

-DISCUSSION-

Any induction motor, single-phase or three-phase, that is connected to the ac bus and is forced to rotate faster than its synchronous speed will send electrical power to the ac bus. An induction motor operating this way is called an induction generator or an asynchronous generator.

The operation of an induction generator is different from that of a synchronous alternator. An induction generator always requires reactive power (lagging vars) to cause rotor current. It requires lagging vars, regardless of the amount of real power (watts) it sends out. A synchronous alternator does not require reactive power to produce watts. This means that a synchronous alternator can be used, by itself, to supply a purely resistive load, while an induction generator cannot, by itself, be used this way.

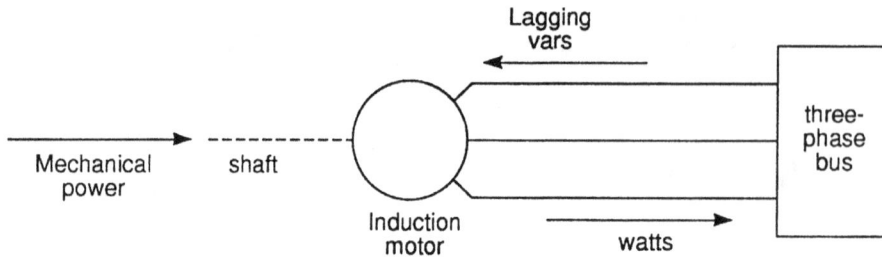

FIGURE 11-1 Induction Generator Connected to the Bus

However, an induction generator can be used to supply a resistive load by placing suitable capacitors across the output of the induction generator. The capacitors supply the lagging vars necessary to excite the rotor.

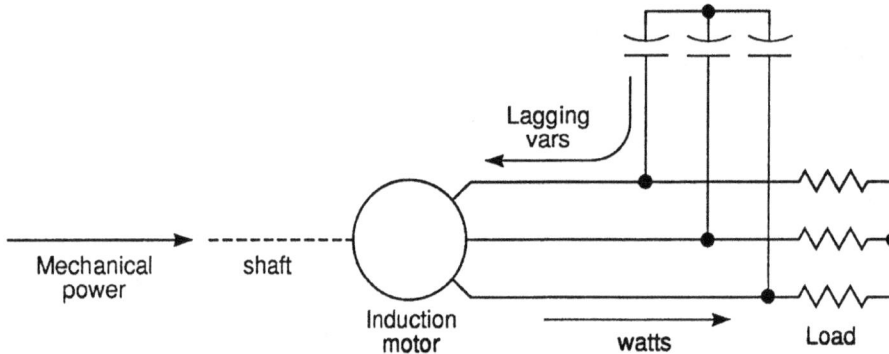

FIGURE 11-2 Induction Generator Connected to a Resistive Load

Because an induction generator requires lagging vars for operation, it contributes to the bus system instability when paralleled to an ac bus. Capacitors can cancel the lagging vars requirement, but they are expensive. Because of these problems, induction generators are rarely used to generate significant amounts of power.

Induction generators sometimes occur in systems where an induction motor's load overhauls the motor and turns the motor into an induction generator. One example is an induction motor crane lowering a heavy load. Other examples are MG (Motor-Generator) sets, induction motor trains, and induction motor conveyor belt systems.

Induction generators can be used in small-scale power generation. For example, an induction generator would work well with a small hydroelectric power plant.

-PROCEDURE-

1. Copy the nameplate data of the three-phase squirrel cage induction motor.
2. Draw a circuit diagram of the dynamometer's dc motor connected in such a way as to be speed-torque controllable. Connect the dc motor's diagram by a dotted line (indicating the shaft) to a circuit diagram of the induction motor connected to the ac bus. Include meters in your diagram as in Experiment 10.
3. Construct the circuits and have them approved by the instructor.
4. Short the current coil of the wattmeter and the ac ammeter.
5. **Turn off all power to the ac motor.** Energize the dc motor and note the direction of rotation of the shaft.
6. **Turn off all dc power to the dc motor.** Energize the ac induction motor and note the direction of rotation of the shaft.
7. If the rotations are the same, go to Procedure 8. If the rotations are not the same, reverse two of the three-phase input leads and repeat Procedure 6. The rotations should then be the same. **It is very important that both motors produce the same rotation.**
8. Make a data table to record each meter reading as the induction generator is loaded in 10 steps from no output to 115 percent of rated output. Consider the induction generator to be at rated output when it produces rated current.
9. Start the ac motor.
10. Using lower voltages at first, start the dc motor and increase the dc motor's output until the ac motor is drawing no power.
11. Run a load test on the induction generator.

-QUESTIONS-

1. Draw a complete diagram of the circuits used, including meters.
2. What would have happened if the rotations were not the same and the dc motor's power was increased in Procedure 10? Explain.
3. Using data from this experiment and from Experiment 10, graph percentage full load torque versus percentage synchronous speed over the full range tested, negative to positive load.
5. How many vars is the induction generator drawing at no-load and at full load?

EXPERIMENT 12

SINGLE-PHASE TO THREE-PHASE INDUCTION PHASE CONVERTER

-OBJECTIVE-

To test a single-phase to three-phase induction phase converter.

-EQUIPMENT-

Three-phase induction motor
Starter cord
Three-phase variable resistance load
3PST switch
Fixed 115 volt single-phase ac supply
Ac clamp-type ammeter
VOM
Wattmeter
Tachometer
SPST switch

-DISCUSSION-

A three-phase induction motor may be used to convert single-phase to three-phase. It is generally used when three-phase is not available but is needed to power some equipment requiring three-phase. A small machine shop with some equipment powered by three-phase induction motors might use an induction phase converter.

Once started, a three-phase induction motor can use single-phase to operate as a single-phase induction motor. Its rotation can be explained by the same crossfield theory as that used on single-phase induction motors. The squirrel cage rotor has currents induced into it the same way that they are induced into a single-phase motor rotor. The rotor currents produce rotating magnetic poles that induce voltage into the stator coils not connected to the input voltage.

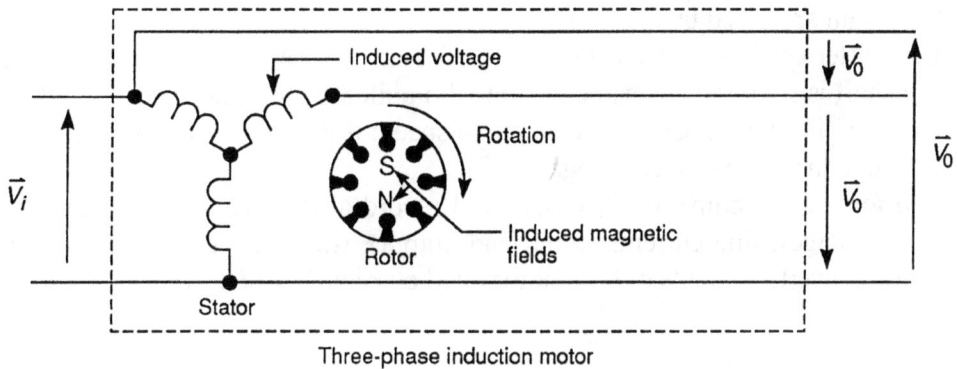

FIGURE 12-1 Three-Phase Induction Motor as an Induction Generator

The two induced voltages from the terminal not connected to the single-phase voltage source relative to the two connected terminals will be almost the same as three-phase line to line voltages. The symmetrical physical construction of three-phase motors creates the correct phase sequences. The only difference is that the two induced voltages are slightly less in magnitude than those of the single-phase voltage source, due to stator winding voltage drops.

A Delta wound three-phase induction motor works equally well as a Wye wound as a single-phase to three-phase converter.

-PROCEDURE-

1. Construct the circuit of Figure 12-2 with the R's disconnected and the wattmeter's current coil short-circuited.

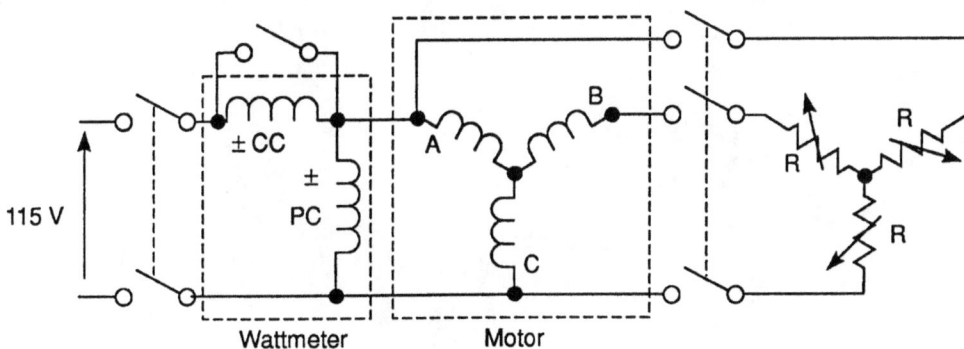

FIGURE 12-2

2. Have the circuit approved by the instructor.
3. Wrap the starter cord around the motor's shaft and pull on it to start the shaft spinning. When the shaft is spinning, the same person who pulled on the starter cord should close the circuit breaker. The motor should accelerate to speed; if it does not, open the circuit breaker and go through the same procedure again.
4. Do a load test on the converter. Measure and record all line voltages (input and output), load phase voltages, line currents (input and output), wattmeter readings, and tachometer readings for converter outputs of 0 to motor rated current. Take 6 to 10 sets of readings.

-QUESTIONS-

1. Make a graph of percentage efficiency (three-phase output power over single-phase input power) and percentage power factor (as seen at the single-phase input) versus three-phase output power.
2. How unbalanced was your three-phase? Where might this amount of imbalance be a problem?
3. A three-phase motor induction converter can convert enough single-phase to three-phase to start an unloaded three-phase induction motor of about twice its size. However, if the same twice sized three-phase induction motor is loaded down, the induction converter won't be able to start it. Explain why?

EXPERIMENT 13

NO-LOAD ALTERNATOR
SATURATION CURVE

-OBJECTIVE-

To demonstrate the relationship between the no-load voltage and the dc field current of an alternator.

-EQUIPMENT-

Alternator, single-phase or three-phase
Dc motor
3 variable 115 volt dc supplies
Dc ammeter
2 dc voltmeters
VOM
Oscilloscope
Tachometer

-DISCUSSION-

Alternators produce all significant amounts of electrical power today. They are the source of electric power from all electric utility power stations and will continue to be so for the foreseeable future.

Saturation occurs in a magnetic material when the material has been magnetized to nearly the highest possible degree. This experiment shows that the steel of an alternator's field approaches saturation when excess current is put through the alternator's field winding. This will be seen in the decreasing output, V_O, to field current, I_F, ratio as I_F increases.

66

A useful equation for working with alternators is:

$$f = P \times S/120$$

where

    f = alternator output frequency (Hz)
    P = number of alternator poles
    S = speed of alternator (rpm)

-PROCEDURE-

1. Draw a circuit diagram of a separately-excited dc motor. Link that circuit diagram with a dotted line to a circuit diagram of an ac alternator. On the dc motor circuit, include meters that measure armature voltage and current. On the ac alternator circuit, include a meter to measure input dc voltage and an oscilloscope to measure output ac voltage and frequency.
2. Record all alternator nameplate data.
3. Construct the circuits.
4. Have the circuits approved by the instructor.
5. Run the alternator at rated speed.
6. Starting with the alternator field set at 0 volts, increase field voltage in 10 steps to 1.5 times rated field current. Then decrease the field, through the same field voltage steps, back to 0 volts. Tabulate field voltage, field current, frequency, and output voltage. Use one line-to-line voltage as the alternator's output voltage, if using a three-phase alternator. Be certain to keep rpm constant at rated rpm throughout this step.

-QUESTIONS-

1. Plot the ascending and descending curves of alternator output voltage versus field current.
2. Is the descending curve higher than the ascending curve? If so, why?
3. What is the relationship between the number of poles, frequency, and rpm for an alternator?
4. Why is it important to keep the rotor speed constant in this experiment?
5. What is the use of nameplate data? What might happen if a machine were to be used at voltages and currents above its nameplate rating?

EXPERIMENT 14

ALTERNATOR LOAD CHARACTERISTICS

-OBJECTIVE-

To compare the calculated load characteristics of an alternator to those found by actually testing of the alternator under load.

-EQUIPMENT-

Alternator, single-phase or three-phase
Dc motor
Variable resistance load, single-phase or three-phase, depending on the alternator
3 variable 115 volt dc supplies
Ac ammeter
Ac voltmeter
Dc ammeter
2 dc voltmeters
VOM
Tachometer

-DISCUSSION-

Equivalent Circuits

An alternator can be thought of as an ac source(s) in series with an equivalent impedance(s). This is seen in the following three figures.

68

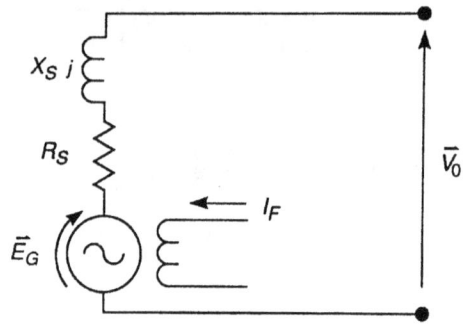

FIGURE 14-1 Single-Phase Alternator Equivalent Circuit

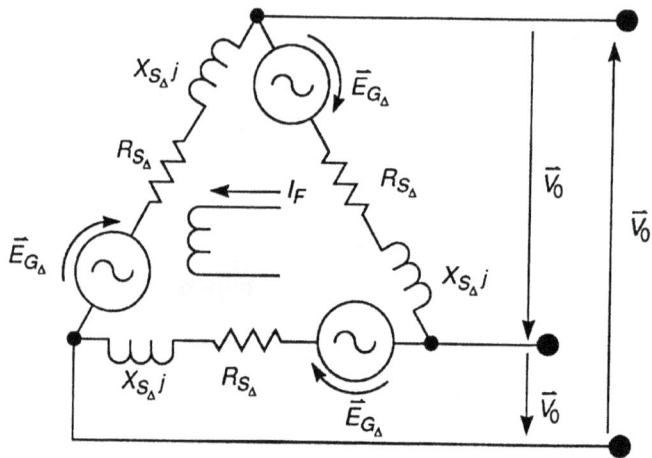

FIGURE 14-2 Three-Phase Delta Wound Alternator Equivalent Circuit

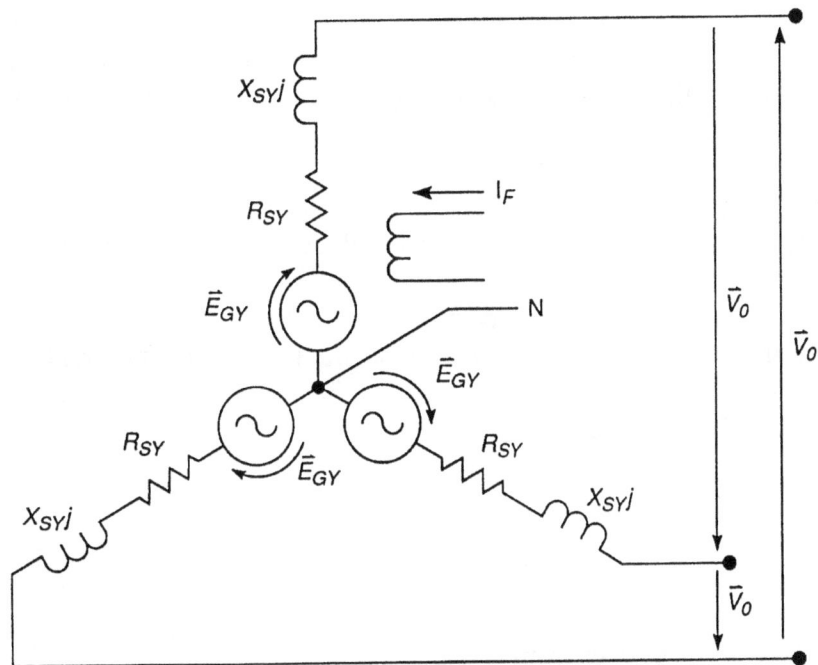

FIGURE 14-3 Three-Phase Wye Wound Alternator Equivalent Circuit

Determining R_S, X_S, and E_G

Knowledge of the values of R_S, X_S, and E_G can be useful in solving alternator circuit problems.

To determine R_S:
  a. For a single-phase alternator, connect a low dc voltage to the non-rotating alternator's output terminals and apply rated current.

$$R_S = 1.5 \times V/I$$

(*Note*: The 1.5 factor approximates the amount the effective ac resistance is greater than the dc resistance.)

  b. For a three-phase Wye or Delta alternator, connect a low dc voltage to two of the non-rotating alternator's ac output terminals and apply rated current.

$$R_{SY} = 0.75 \times V/I$$

(*Note*: It is often easier to do circuit calculations with three-phase alternators represented by their Wye equivalents. Only Wye equivalents will be used here, regardless of a three-phase alternator's actual type. All circuit calculations can be done correctly here with a Delta wound alternator represented as its Wye equivalent.)

To determine Z_S:

    a. Short the output(s) of the alternator. Rotate it at rated speed and apply enough field current to have the alternator produce rated output line current. Record the field current value.

    b. Disconnect the alternator output short-circuit. Again drive the alternator at rated speed with the same field current as in 1. Record the line-to-line output current.

    c. Calculate Z_S:

      i. For a single-phase alternator:

$$Z_S = V_O/I_O$$

      where

V_O = alternator output voltage (rms volts)
I_O = alternator output rated current (rms amps)

      ii. For a three-phase Wye or Delta alternator:

$$Z_{SY} = V_O/(\sqrt{3} \times I_R)$$

      where

V_O = line-to-line alternator output voltage (rms volts)
I_R = rated output alternator line current (rms amps)

To determine X_S:

    a. For a single-phase alternator:

$$X_S = (Z_S^2 - R_S^2)^{.5}$$

    b. For a three-phase Wye or Delta alternator:

$$X_{SY} = (Z_{SY}^2 - R_{SY}^2)^{.5}$$

To determine E_G:

    Run the alternator at rated speed and rated field current without load.

    a. For a single-phase alternator:

$$E_G = V_O$$

b. For a three-phase Wye or Delta alternator:

$$E_{GY} = V_O/\sqrt{3}$$

-PROCEDURE-

1. Draw a circuit for determining R_S. Include meters in the circuit.
2. Construct the circuit and determine R_S.
3. Draw the circuits for determining Z_S or Z_{SY}. Include meters in your circuit.
4. Construct the circuit, have it approved by the instructor, and determine Z_S or Z_{SY}.
5. Run the alternator with rated field current at rated speed with no load to determine E_G.
6. Design and draw a load circuit (Wye if three-phase) for running a load test on the alternator.
7. Construct the circuit and have it approved by the instructor.
8. Setting the field current to its rated value and being careful to maintain alternator rated speed, run a load test. Record output voltage versus current for 10 steps up to 1.25 times rated current.

-QUESTIONS-

1. Determine R_S, X_S, and E_G. Show these values in an equivalent circuit of the alternator (Wye if three-phase).
2. Using the equivalent alternator circuit, calculate points for a graph of alternator output voltage versus output current for a resistive load. Plot these points from 0 to 125 percent rated output current.
3. Superimpose the data of the load test of Procedure 8 onto the curve just plotted. Are the curves similar? Account for the differences.
4. What are the advantages and disadvantages of calculating alternator load characteristics, as opposed to running actual alternator load tests?

72

EXPERIMENT 15

PARALLELING AN ALTERNATOR WITH AN INFINITE BUS

-OBJECTIVE-

To parallel an alternator with the 60 Hz bus and to observe the alternator's behavior as it is loaded.

-EQUIPMENT-

Dc motor
3 variable 115 volt dc supplies
Ac ammeter
2 ac voltmeters
Dc ammeter
2 dc voltmeters
VOM
Tachometer
Either A or B
    A. Single-phase alternator
       Light bulb
       Light bulb socket
       1PST switch
       Fixed 115 volt single-phase ac bus
       Wattmeter
    B. Three-phase alternator
       3 light bulbs
       3 light bulb sockets
       3PST switch
       Fixed 115 volt three-phase ac bus
       3 wattmeters

-DISCUSSION-

The thousands of electric power company alternators paralleled together in the U.S.-Canada electric grid are in effect an infinite bus. Both a student connecting an alternator's output to a laboratory outlet and a power company connecting an 800 mva alternator to the grid will observe similar problems in synchronizing and similar alternator characteristics.

Alternator Paralleled with an Infinite Bus

The operator of an alternator paralleled with an infinite bus can adjust the field current and the input shaft torque. The alternator speed, output frequency, and phasor output voltage will be constant as determined by the frequency and phasor voltage of the infinite bus.

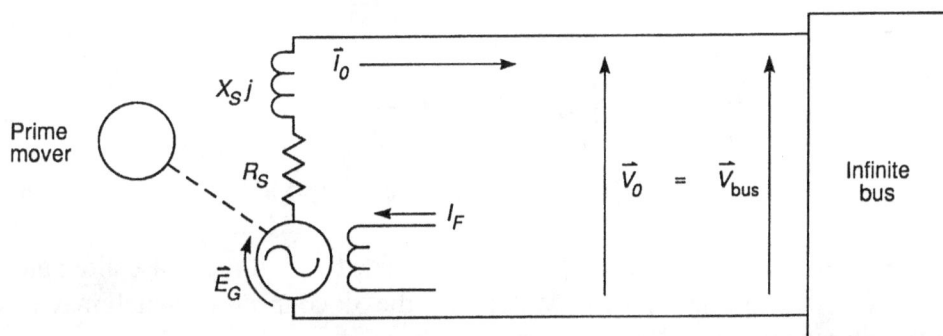

FIGURE 15-1 Single-Phase Alternator Paralleled with an Infinite Bus

Increasing the field current will not increase the power out of an alternator paralleled with an infinite bus. A change of field current will increase or decrease the output current of an alternator but will, simultaneously, cause the phase angle, Θ, between V_O and I_O to increase or decrease, respectively. The resultant alternator output power remains constant. Output power is $V_O \times I_O \times \cos\Theta$ for a single-phase alternator and $\sqrt{3}V_O \times I_O \times \cos\Theta$ for a three-phase alternator.

Increasing the input shaft torque to an alternator will increase the power out of an alternator paralleled with an infinite bus. As the shaft torque of an alternator is increased, the phase angle, , between the I_O and V_O becomes less and I_O becomes greater. This causes the alternator output power = $V_O \times I_O \times \cos\Theta$ to increase. There is a maximum allowable shaft torque for each field current. Beyond this torque, the alternator would lose synchronism with the infinite bus, and large undesirable currents would pass between the alternator and the infinite bus. Hopefully, a circuit breaker would prevent the destruction of the alternator.

Dark Lamp Method

There are several methods of determining when an alternator is synchronized with an infinite bus or another alternator. The simplest of these is the dark lamp method.

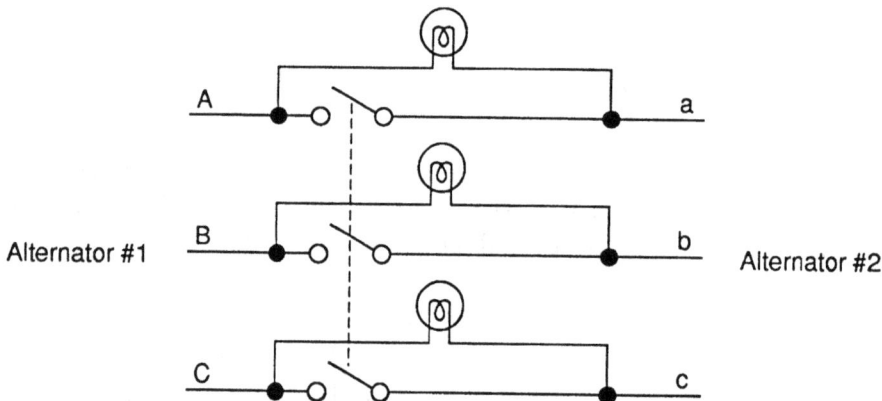

FIGURE 15-2 Three-Phase Dark Lamp Method

When three-phase alternator #1 is properly synchronized with three-phase alternator #2, all three lamps will be dark simultaneously. At this time the 3P connecting switch may be closed, paralleling the alternators.

The dark lamp method is the same for paralleling single-phase alternators, except that only one lamp and a 1P connecting switch are required.

-PROCEDURE-

1. Draw a diagram of the circuit required to run the dc motor and the circuit of the alternator ready to be connected to the bus via the dark lamp method.
   Include in your diagram:
   a. A dc voltmeter and ammeter to determine dc motor armature input power.
   b. An ac voltmeter and ac ammeter to measure alternator output voltage and output current.
   c. An ac voltmeter to measure bus voltage.
   d. A dc voltmeter to measure alternator field voltage.
   e. A wattmeter to measure alternator output power (3 wattmeters if three-phase).
2. Have your circuit diagram approved by the instructor.
3. Measure the alternator's field resistance.
4. Construct the circuit. **Be sure to leave the connection switch open.**
5. Have the circuit approved by the instructor.

6. Run the alternator up to synchronous speed and bus voltage. Record all meter readings. How is the light bulb(s) lighting?
7. If a three-phase alternator is being paralleled, stop it and reverse two of its output leads. Run the machine up to the same meter readings as recorded in Procedure 6. How are the light bulbs lighting?
8. Have the instructor observe the synchronization and paralleling of the alternator.
9. Increase the armature current to the dc motor from that seen at synchronization by about 5 percent. Then without further adjusting the dc motor's voltages, decrease the alternator field current in five steps until the alternator produces rated output current. Record all meter readings at each step.
10. Leave the dc motor running as it is. Increase the alternator field current back to where it initially was in Procedure 9. Then, increase the alternator field current in five steps until it again produces rated output current. Record all meter readings at each step.
11. Leave the alternator field current at the final value of Procedure 10. Increase the power to the dc motor in five steps, so that the alternator output is increased to about one-half of its rated output power. Record all meter readings at each step.

-QUESTIONS-

1. Draw the circuit diagram of the dc motor-alternator circuit and its connection to the bus via the dark lamp method. Include all meters in the diagram.
2. What would be indicated if the following was observed in a three-phase alternator's dark lamp test before paralleling? Explain your answers briefly.
   a. All the lights slowly go dim and then slowly go bright in a repeating cycle.
   b. One light goes out, while the other two are bright.
   c. All lights are steadily dimly lit.
3. Draw one graph of alternator output current versus field current from the data of Procedures 9 and 10.
4. Did the wattmeters record a change in alternator output power in Procedures 9 and 10? Explain.
5. Draw a graph of power out of the alternator versus the power into the dc motor from the data of Procedure 11.
6. Draw a graph of the alternator power factor versus the output power of the alternator from the data of Procedure 11.

EXPERIMENT 16

SINGLE-PHASE CAPACITOR-START INDUCTION MOTOR CHARACTERISTICS

-OBJECTIVE-

To determine the torque versus speed curve of a single-phase capacitor-start induction motor and to determine the starting capacitance needed for maximum starting torque.

-EQUIPMENT-

Single-phase capacitor-start induction motor
3 motor starting capacitors
Variable 115 volt single-phase ac supply
Variable 115 volt dc supply
Dynamometer with blocked rotor capability
Variable resistance load
Ac ammeter
Ac voltmeter
Dc ammeter
Dc voltmeter
VOM
Tachometer
SPST switch

-DISCUSSION-

Single-Phase Capacitor-Start Motor During Starting

During starting, a capacitor-start motor has two separate stator winding circuits connected to the single-phase source, a starting circuit, and a running circuit. The centrifugal starting switch is closed at this time.

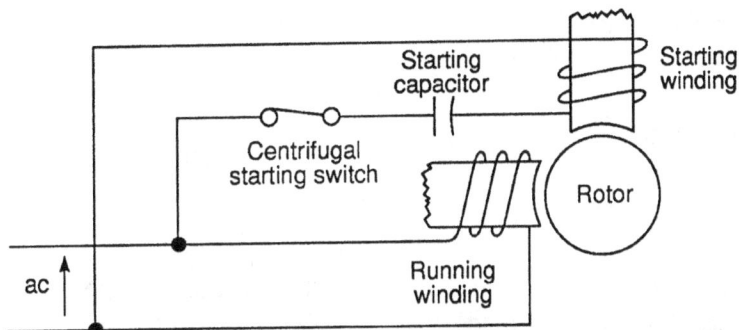

FIGURE 16-1 Single-Phase Motor Equivalent Circuit During Starting

During starting, currents are flowing through both the running winding and the starting winding. Because of the capacitance in the starting winding circuit, the current through the starting winding leads the current through the running winding. The leading current in the starting winding causes the magnetic field produced by the starting winding to occur before that produced by the running winding. The out-of-phase magnetic fields then combine to produce a rotating field.

The strength of the starting rotating field depends on the field strengths of the starting and running windings and the phase relationship between them. The strength of each winding's field depends on the magnitude of current through it.

For a given running winding and starting winding at a given supply voltage, the strength of the starting rotating field depends mainly on the phase relationship between the produced magnetic fields. If the fields are in phase, there will be no rotating field; if the fields are 90 electrical degrees (one-fourth of an input voltage waveform period) out of phase, there will be a maximum rotating field; if the fields are 180 electrical degrees out of phase there will be no rotating field. The phase relationship between the magnetic fields is set by the value of the starting capacitor.

Equations That Will Determine the Starting Capacitance Value for Maximum Starting Torque

R_{SW} = ac resistance of the starting winding, equals 1.25 times its dc resistance (ohms)
X_{SW} = inductive impedance of the starting winding (ohms)
X_C = capacitive impedance of the starting capacitor (ohms)
$\mathbf{Z}_{SC}$ = phasor impedance of the starting circuit, starting winding, and starting capacitor
    = $R_{SW} + (X_{SW} - X_C)j$ (ohms)

78

R_{RW} = ac resistance of the running winding, equals 1.25 times its dc resistance (ohms)

X_{RW} = inductive impedance of the running winding (ohms)

Z_{RW} = phasor impedance of the running winding, running winding only

$\quad = R_{RW} + X_{RW}j$ (ohms)

K $\quad$ = some unknown constant

For near maximum starting torque:

$$\mathbf{Z}_{SC} \times K\angle 90° = \mathbf{Z}_{RW}$$

For a motor where the running winding and starting winding impedances are known, it is possible to use the above equation to solve for X_C. The alert reader might ask how it is possible to solve one equation that has two unknowns, K and X_C. It can be solved, because the single equation actually is two equations: one equation of real numbers and one equation of imaginary numbers. All that need to be done is to substitute all the known values into the equation, separate the real numbers from the imaginary numbers, and then write one equation for the real and one for the imaginary. The equations can then be solved as second-order simultaneous equations.

Single-Phase Capacitor-Start Motor After Starting

After a single-phase capacitor-start motor is up to approximately 75 percent of its synchronous speed, the centrifugal switch opens and disconnects the starting winding from the circuit.

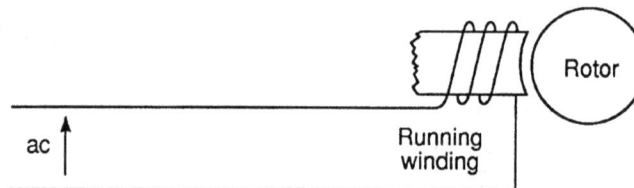

FIGURE 16-2 Single-Phase Motor Equivalent Circuit When at Speed

The rotating rotor produces a field that combines with that of the running winding to produce a rotating field. The rotating field then pulls the rotor around. See an electrical machinery text for a more detailed explanation.

-PROCEDURE-

1. Copy the nameplate data of the single-phase capacitor-start induction motor.

2. Draw a circuit diagram of the motor powered by a single-phase source. Connect the motor's diagram by a dotted line (indicating the shaft) to a circuit diagram of the dynamometer. Include meters in your diagram. Also include a switch for shorting the ac ammeter during motor starting.
3. Construct the circuit and have it approved by the instructor.
4. Short the ammeter.
5. Do a load test on the motor. Record each meter reading as the motor is loaded in 10 steps from no-load to 115 percent of rated load. Consider the motor to be at rated load when it is drawing rated current.
6. Block the rotor with the rotor block so that the blocked (non-rotating) rotor torque can be measured. Short the ammeter. Have the setup approved by the instructor.
7. Apply half voltage to the motor for an instant and quickly measure the produced torque. Record the torque in Table 16-1. **Do not leave the power on for more than a few seconds; doing so could destroy the starting capacitor or the motor windings.**

TABLE 16-1 Motor Torque Versus Starting Capacitance

C (μF)	Measured T at ½ voltage (ft lb)	Calculated T at full voltage = 4 x (T at ½ v) (ft lb)
motor's C =		
approx. 2 x motor's C =		
approx. 3 x motor's C =		
approx. ½ x motor's C =		
approx. 1/3 x motor's C =		

8. Connect different starting capacitances into the motor's starting circuit. Test for blocked rotor torque at each capacitance. Create the different capacitances by connecting capacitors in series or parallel.
9. Measure the dc resistance of the starting winding and the running winding with an ohmmeter. Multiply each by 1.25 to get their approximate resistances at 60 Hz.
10. Apply a low ac 60 Hz voltage to the starting winding and running winding to determine their impedances.

$$Z = V/I$$

11. Find the inductive impedances of the windings by the equation.

$$X = (Z^2 - R^2)^{.5}$$

-QUESTIONS-

1. Draw complete diagrams of the circuits used, including meters.
2. Draw a graph of percentage full load torque versus percentage synchronous speed. Include the result of the blocked rotor test with the motor's capacitor on the graph and estimate the region in between. Don't forget that there is a sudden decrease in torque at about 75 percent of synchronous speed when the centrifugal switch disconnects the starting circuit.
3. Draw a graph of produced blocked rotor torque versus starting capacitance. What capacitance value would produce the maximum starting torque? How different is this from that of the motor's capacitor?
4. Using the data from Procedures 9 to 11, calculate the capacitance that should produce the maximum starting torque. How close is this figure to that determined by the blocked rotor torque versus starting capacitance curve? Explain.

PART IV

MOTOR CONTROL EXPERIMENTS

RELAY-CONTACTOR CONTROL

EXPERIMENT 17

DEFINITE TIME ACCELERATION DC MOTOR STARTER

-OBJECTIVE-

To construct and operate a definite time acceleration dc motor starter.

-EQUIPMENT-

Dc motor
Time delay relay
Contactor
Slidewire rheostat
Starting resistance
DPST knife switch
Fixed 115 volt single-phase ac supply
Fixed 115 volt dc supply
Dc ammeter
VOM
Stopwatch

-DISCUSSION-

Modern installations don't use motors with manual starters like those of Experiment 8 because it is too easy to forget to open or close the starting resistance switch. Instead, an automatic system is used to gradually bring a motor to full line voltage. This can be done with a time acceleration starter which, following a timed sequence, shorts out starting resistors until a motor is directly connected to the line.

In this experiment, a single time delay relay is energized when dc is first connected (manually in this experiment) to the shunt motor. After a period of time, the time delay activates and closes contacts to operate a contactor that shorts out the starting resistor. The motor is thus connected directly to the line.

Relays

A relay is an electrically operated switch. When voltage is applied to a relay's coil, a magnetic field is produced to move an iron core that mechanically opens or closes electrical contacts. Typical voltage ratings for relay coils are between 12 and 115 volts. Usually, the current rating of a relay's contacts are a few amperes or less.

In circuit diagrams, relay coils are represented by circles, and relay contacts are represented by parallel lines. Examples are shown in Figure 17-1. A relay's coil is designated by the same letters as its contacts. In Figure 17-1 voltage applied to the A coil will close the A contacts.

FIGURE 17-1 Symbols for a Relay Coil and Contacts

Time Delay Relays

This experiment uses a time delay relay. A time delay relay is a special relay that has a set time delay between the time its coil receives voltage and its contacts close. Time delay relays use the letters TD as part of their designation

The time delay relay of this experiment is mechanical. The coil creates a magnetic field to pull its iron core as in an ordinary relay, but the iron core must pull a piston through an oil-filled cylinder. The viscosity of the oil slows the motion of the piston, which delays the closing of the contacts.

Contactors

Contactors are simply large relays. They switch larger voltages and currents. Their sizes range from that of a book to that of an automobile. Smaller contactors are used to handle motor or lighting loads. The largest contactors are in power company switchyards and can handle thousands of volts and amps. In schematics they are represented with the same symbols as relays.

-PROCEDURE-

1. Determine the current versus time characteristics curve of the time delay relay. Do this by applying a range of currents to the relay's coil and by recording the time it takes for the relay to activate. Be sure to allow the relay to fully reset between readings. Do not exceed the current rating of the coil. Plot the data on the graph of Figure 17-2.

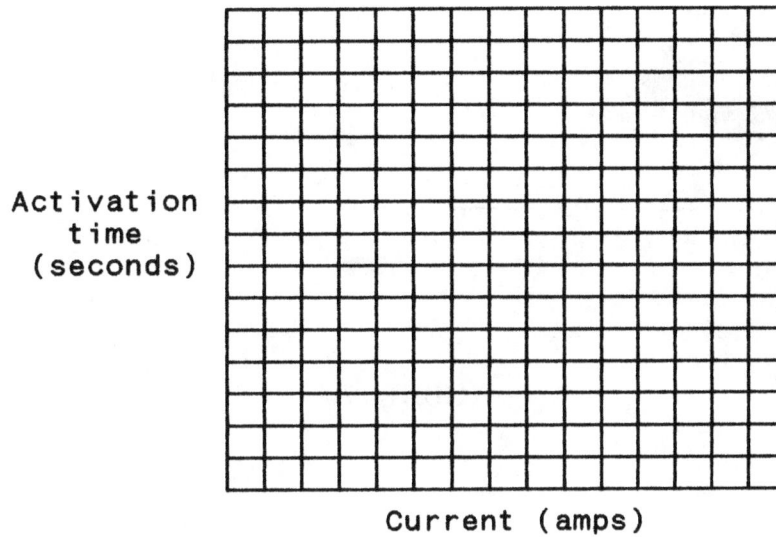

FIGURE 17-2

2. Construct the dc shunt motor circuit of Figure 17-3 without constructing the circuits of Figures 17-4 and 17-5. Choose a starting resistance, R_S, so that motor current at starting is less than or equal to the motor's rated full load current.

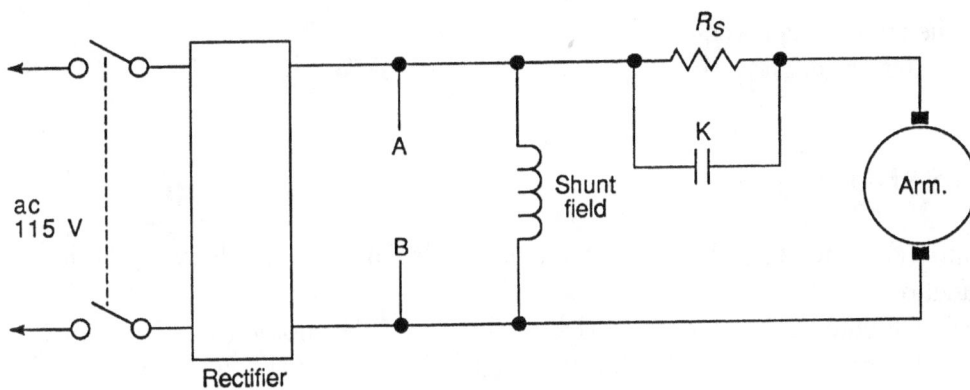

FIGURE 17-3

3. Have the circuit approved by the instructor.
4. Start the motor several times and measure the time it takes to come to a steady speed.
5. Using the curve drawn on Figure 17-2 and calculations, determine the R required to be put in series with the time delay relay's coil so that the starting resistance is shorted when the motor reaches its steady speed.

6. Construct the circuits of Figures 17-4 and 17-5.

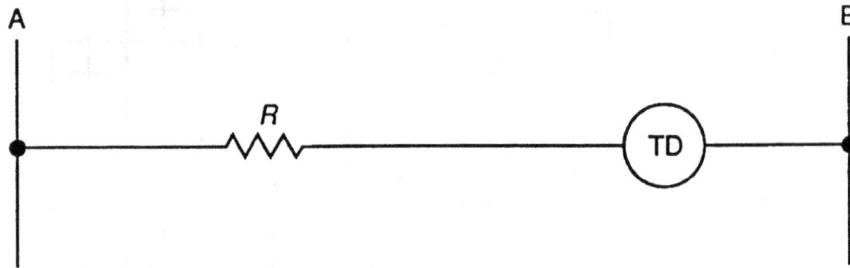

FIGURE 17-4

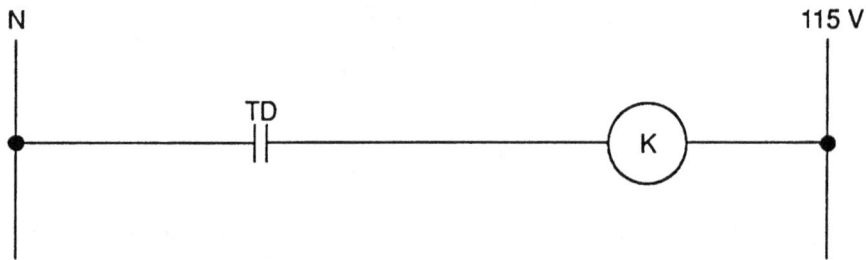

FIGURE 17-5

7. Have the circuits approved by the instructor.
8. Operate the circuits.

-QUESTIONS-

1. Explain how the time delay relay operates. Use a physical drawing of it with your explanation.
2. Redesign Figure 17-3 using a 4PDT switch so that the motor can be run in forward or reverse. Increase the size of R_S and R so that an operator may suddenly move the switch from forward to reverse. Show all calculations and a circuit diagram.

EXPERIMENT 18

RELAY-CONTACTOR MOTOR CONTROL

-OBJECTIVE-

To construct a control circuit that will run a three-phase induction motor in forward and reverse.

-EQUIPMENT-

Three-phase squirrel cage induction motor
2 relays
Three-phase circuit breaker
2 three-phase contactors
3 pushbuttons
2 lamp bulbs
Fixed 115 volt three-phase ac supply
VOM

88

-DISCUSSION-

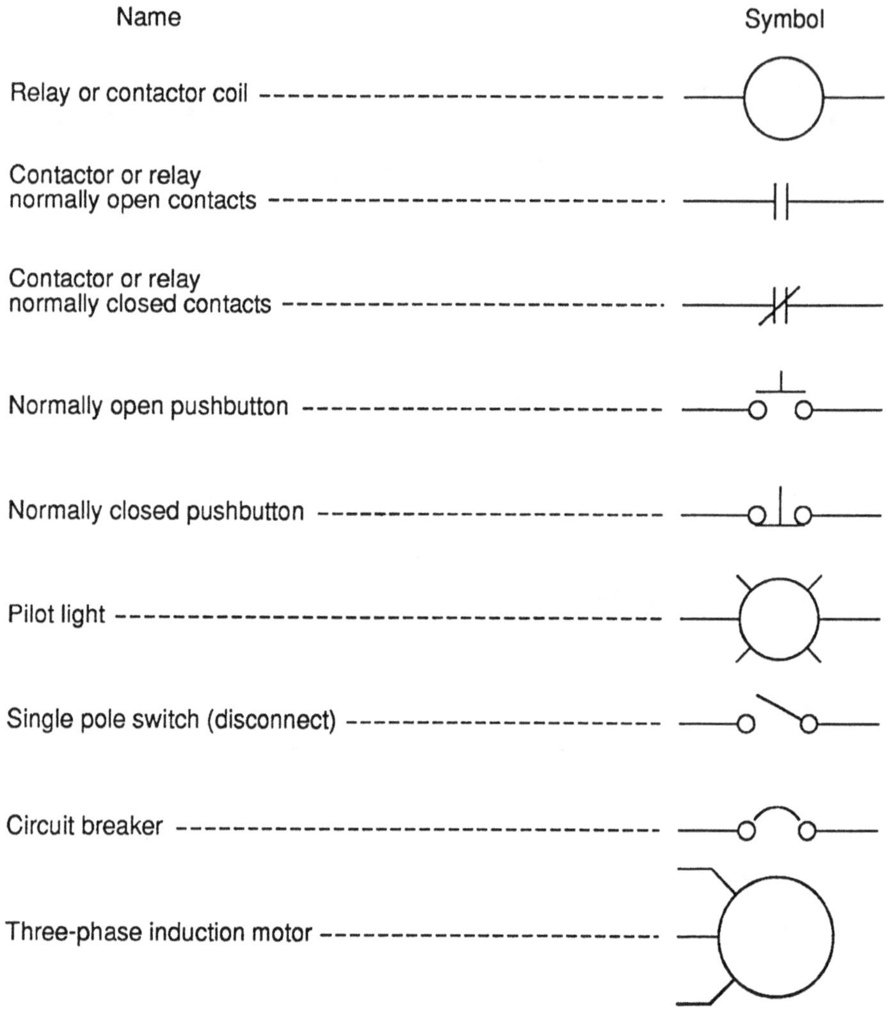

Name	Symbol
Relay or contactor coil	○
Contactor or relay normally open contacts	‖
Contactor or relay normally closed contacts	⫫
Normally open pushbutton	○ ○
Normally closed pushbutton	○ ⊥ ○
Pilot light	⊗
Single pole switch (disconnect)	○ ／○
Circuit breaker	○ ⌒ ○
Three-phase induction motor	○

FIGURE 18-1 Some Graphic Symbols for Motor Control Circuits

-PROCEDURE-

1. Answer the following questions on a sheet of paper before going to Procedure 2.
   a. What are the voltage and current ratings of your relays and contactors?
   b. What is your motor's full load current?
   c. What is the rating of the supplied circuit breaker? Is it the right size?
   d. What is the worst mistake that could be made in connecting these circuits?

2. Construct the circuits of Figures 18-2 and 18-3.

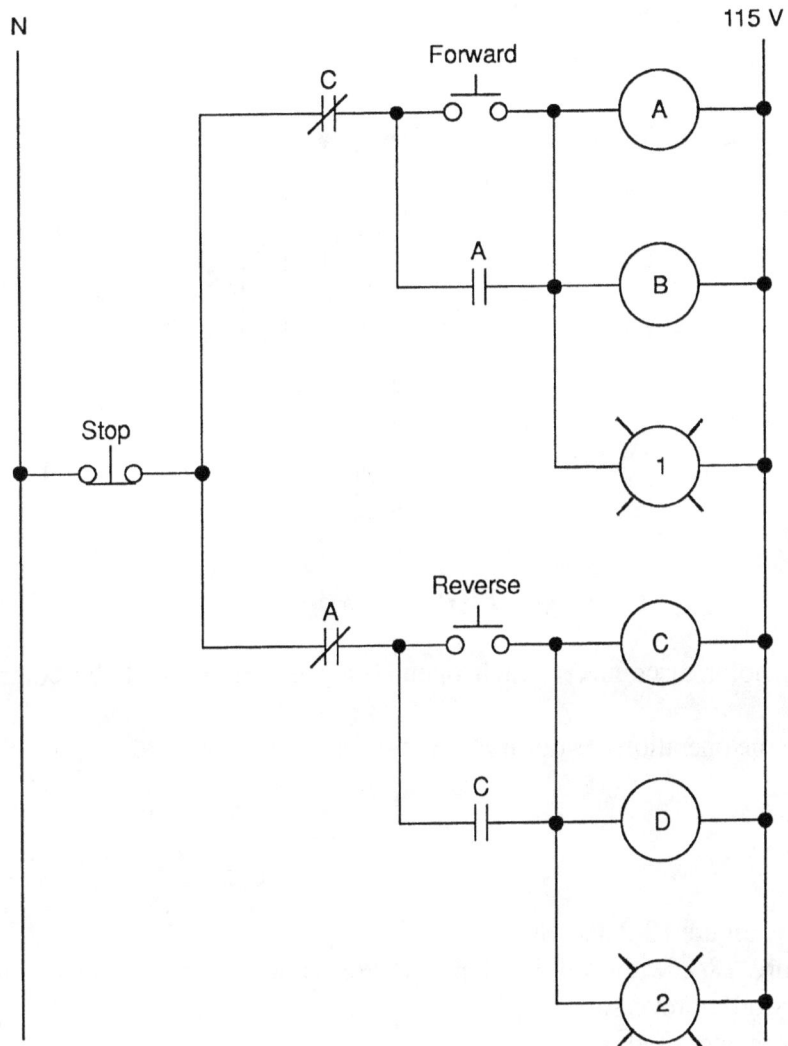

FIGURE 18-2

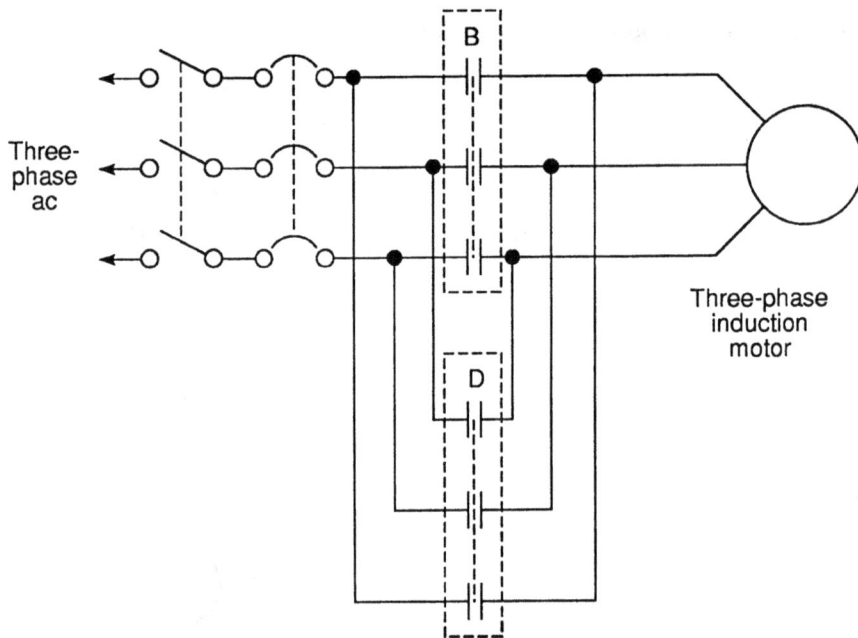

FIGURE 18-3

3. Leaving the motor disconnect switch open, test the operation of the control circuit and contactors.
4. Demonstrate the operation of your motor controller to the instructor.

-QUESTIONS-

1. Neatly redraw Figure 18-2 and 18-3.
2. Redraw Figure 18-2 with additional pushbuttons and relays so that the motor may be jogged in forward or reverse.
3. Where is a contactor used on a car?
4. List some uses of relays, contactors, or limit switches that you have seen outside the laboratory.

EXPERIMENT 19

AUTOMATIC DYNAMIC BRAKING
OF INDUCTION MOTORS

-OBJECTIVE-

To construct a control circuit that will automatically dynamically brake a three-phase induction motor.

-EQUIPMENT-

Three-phase induction motor
Time delay relay
Slidewire rheostat
3 relays
Three-phase circuit breaker
2 three-phase contactors
2 pushbuttons
Transformer
Bridge rectifier
Fixed 115 volt three-phase ac supply
VOM

-DISCUSSION-

Any induction motor can be brought to a quick stop by applying dc to its stator winding. On a three-phase induction motor any two terminals may be used.

Braking action takes place as the rotor bars cut the dc flux of the stator. Rotational energy is converted to current that is then dissipated as heat in the rotor by $I^2 \times R$ heating.

Typically applied dc voltages are between one-tenth and one-third of rated ac voltages. A dc voltage of a magnitude of one-third of the rated will produce an overcurrent in the stator, but this won't harm the stator since it is only applied temporarily.

-PROCEDURE-

1. If necessary, refer to Experiment 18 for a discussion of time delay (TD) relays.
2. Construct the circuit of Figure 19-1. Test the circuit for point-to-point continuity with a VOM.

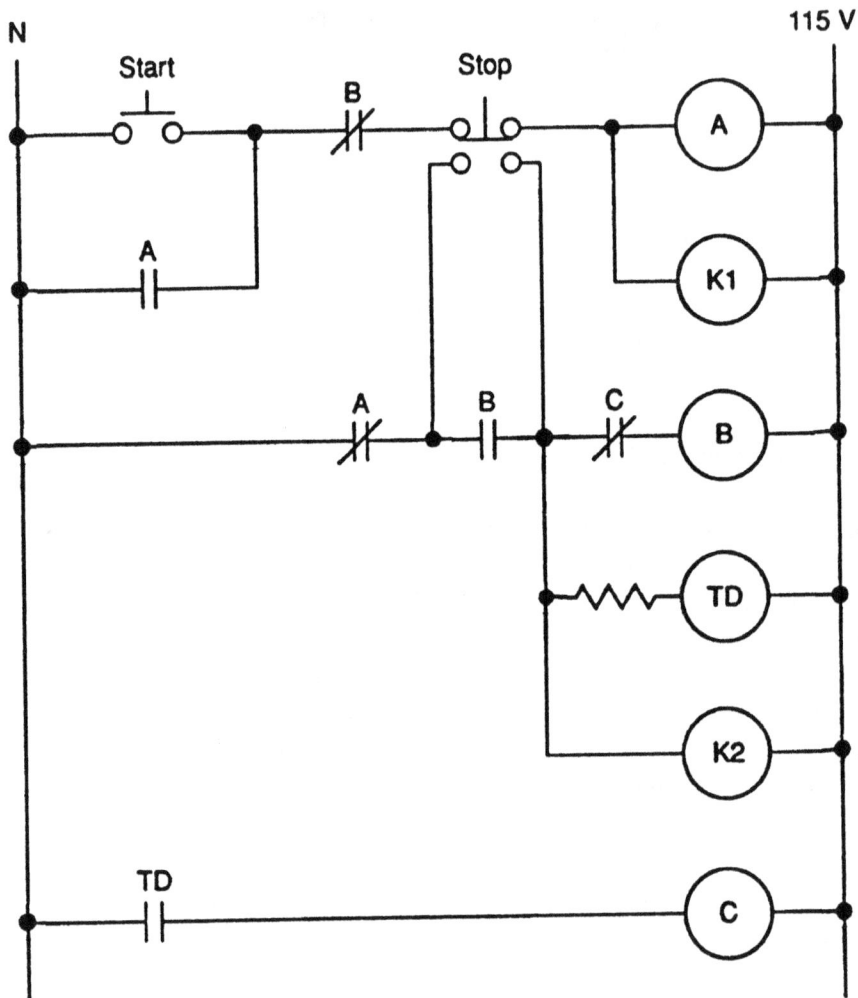

FIGURE 19-1

3. Set the TD to a 3-second delay, for initial testing. Set the delay by adjusting the TD's plunger position or by changing the amount of resistance in series with the TD's coil.
4. Construct the circuit of Figure 19-2. Test the circuit for point-to-point continuity with a VOM.

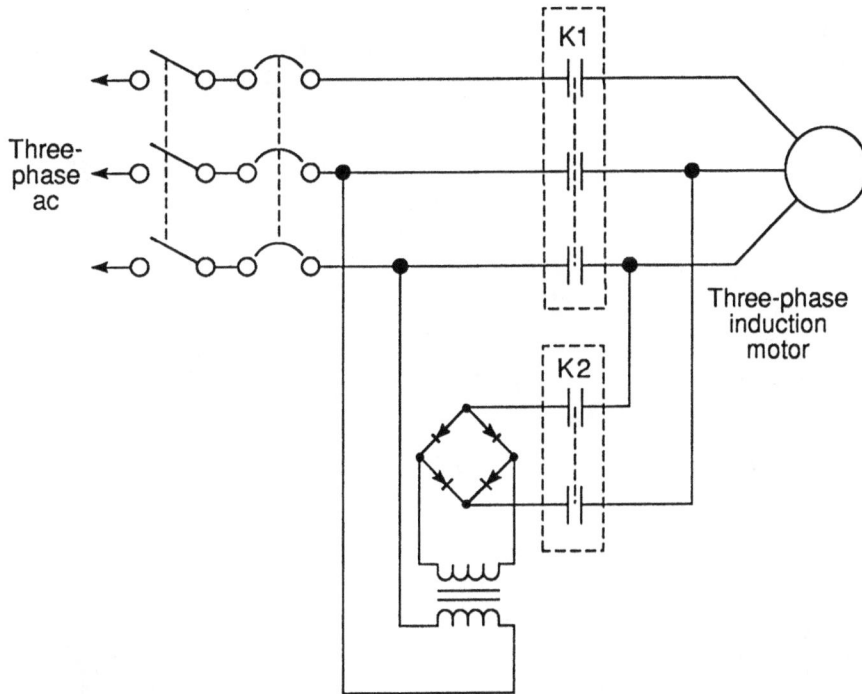

FIGURE 19-2

5. Operate the circuits without three-phase being connected. Check for proper operation.
6. Apply three-phase and operate the circuits.
7. Adjust the TD so that the dc stays on just long enough to stop the motor.
8. Demonstrate the system's operation to the instructor.
9. No written report is necessary for this experiment.

POWER ELECTRONIC CONTROL

EXPERIMENT 20

SINGLE-PHASE TRIAC VOLTAGE CONTROLLER

-OBJECTIVE-

To test the operation of a variable single-phase TRIAC voltage controller on a light bulb and a universal motor.

-EQUIPMENT-

TRIAC, GE SC146D
Trigger diode, GE ST-2
10 ohm, 1/2 watt resistor
3.3 K, 1/2 watt resistor
200 K, 2 watt potentiometer
0.068 μF, 200 volt ac capacitor
0.05 μF, 200 volt ac capacitor
100 watt light bulb
Light bulb socket
Universal motor
Fixed 115 volt single-phase ac supply
VOM
Oscilloscope

-DISCUSSION-

TRIACs are used as light dimmers and as speed controllers for small universal motors like those found in electric power tools.

TRIAC's are simple electronic devices that pass or block the flow of current through them, as directed by a small gate voltage. As can be seen in Figure 20-1, they are three terminal devices. The MTs (Main Terminals) conduct the controlled current. The G (Gate) receives small pulses of control voltage to start conduction between the MTs.

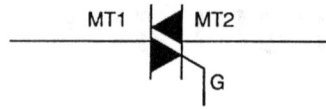

FIGURE 20-1 TRIAC

A TRIAC will:

1. Once in the non-conducting state, stay in the non-conducting state, until both a voltage is applied across its MTs and a G to MT2 voltage pulse is applied.
2. Once in the conduction state, stay in the conduction state, until both the current through its MTs drops to 0 and there is no G to MT2 voltage.
3. Conduct current in either direction through its MTs. TRIACs are designed for use on ac.
4. Be triggered by a gate current of either polarity.

In operation, TRIACs control the effective (rms) voltage by stopping the passage of a current for part of each positive and negative wave. Figure 20-2 shows an ac waveform that has been partially blocked by a TRIAC.

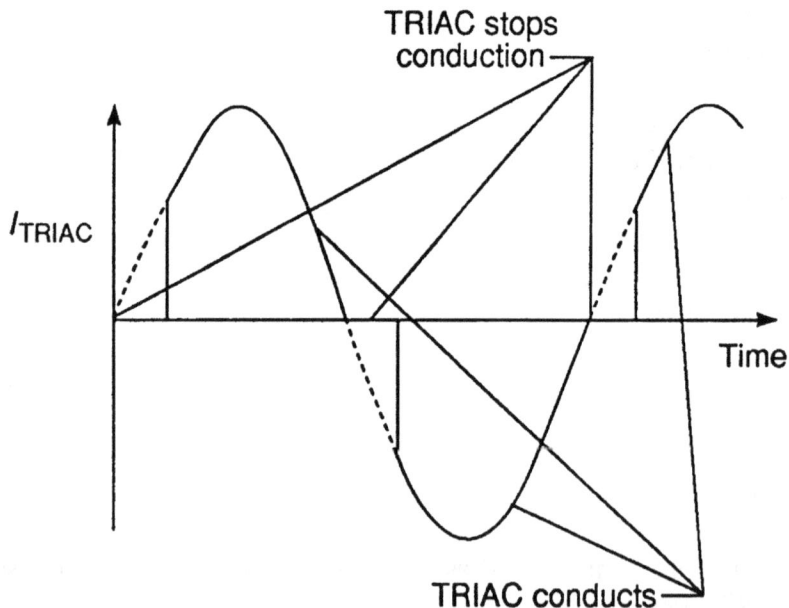

FIGURE 20-2 Typical Current Through a TRIAC

Single-Phase TRIAC Voltage Controller Being Tested

The single-phase TRIAC voltage controller being tested in this experiment is shown in Figure 20-3. In operation, the resistor-capacitor circuit to the left of the TRIAC produces a voltage on the left side of the trigger diode relative to the voltage on the TRIAC's gate. When the voltage difference between those two points exceeds the breakdown voltage of the trigger diode, the trigger diode suddenly conducts to the TRIAC's gate and turns the TRIAC on. Once the TRIAC has turned on, it stays on until the voltage across its MT1 to MT2 drops to 0 volts. Since the ac voltage across it drops to 0 twice every cycle, the TRIAC will only conduct for part of each cycle as was shown in Figure 20-2.

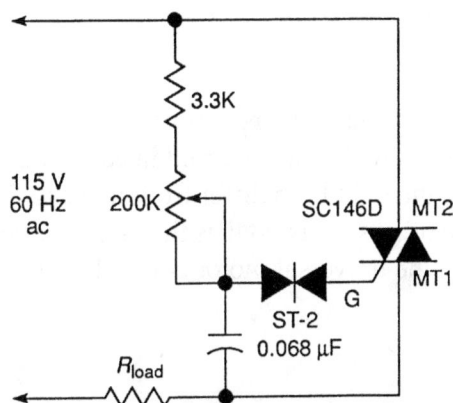

FIGURE 20-3 Voltage Controller

-PROCEDURE-

1. Construct the circuit of Figure 20-3 with the 100 watt light bulb as the load and have the circuit approved by the instructor.
2. Energize the circuit. Vary the potentiometer's setting. It should be possible to adjust the brightness of the bulb. Use the VOM and the oscilloscope to measure the voltage across the bulb for three different potentiometer settings. Don't connect the oscilloscope's ground to the input voltage's live wire; doing so will short the source. To avoid the possibility of shorting the source, connect only the oscilloscope's ungrounded terminal to the circuit. The oscilloscope's grounded plug will complete the circuit. Make sketches of the waveforms.
3. Connect the snubber circuit of Figure 20-4 to MT1 and MT2 of the TRIAC. Replace the 100 watt light bulb with the universal motor. The snubber will protect the TRIAC from the voltage spikes and high dv/dt that the universal motor will produce as its commutator and brushes make and break contact.

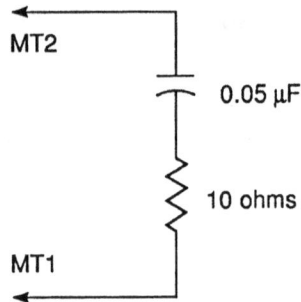

```
MT2 ←─────┐
          │
         ═╪═  0.05 µF
          │
          │
          ⌇  10 ohms
          │
MT1 ←─────┘
```

FIGURE 20-4

-QUESTIONS-

1. Draw the circuit used to control the universal motor.
2. Accurately draw the voltage waveforms observed in each of the tests.
3. How did the waveforms to the light bulb differ from those to the universal motor?
4. Were the waveforms more or less sinusoidal as the effective voltage dropped? Explain.
5. How well was the speed of the universal motor controlled at low voltages? Explain.

EXPERIMENT 21

THREE-PHASE SCR CONVERTER

-OBJECTIVE-

To test the operation of a variable one-quadrant three-phase SCR (Silicon Controlled Rectifier) converter as a power supply for a resistive load and for a dc shunt motor.

-EQUIPMENT-

Variable one-quadrant three-phase SCR converter (already assembled)
4 H inductor
Variable resistance load
Dc motor
Fixed 230 volt three-phase supply
Dynamometer
Ac ammeter
VOM
3 wattmeters (accurately calibrated)
Dual trace oscilloscope (accurately calibrated)
Isolation transformer
10:1 attenuation probe
0.1 ohm shunt resistor

-DISCUSSION-

SCRs

SCRs are the workhorses of the power electronics industry. They range in power capacities from a few watts to thousands of kilowatts. They are similar to TRIACs (used in Experiment 20), although they are simpler in construction and different in operating characteristics.

SCRs pass or block the flow of current through them, as directed by a small gate voltage. As can be seen in Figure 21-1, they are three terminal devices. The anode (A) and the cathode (K) conduct the controlled current. The gate (G) receives small pulses of control voltage relative to the cathode to start conduction from the anode to the cathode.

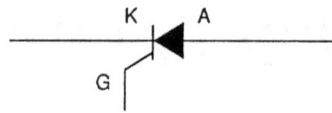

FIGURE 21-1 SCR

A SCR will:

1. Once in the nonconducting state, stay in the nonconducting state, until a positive voltage is applied from its anode to cathode and a gate to cathode voltage is applied.
2. Once in the conduction state, stay in the conduction state, until current from its anode to cathode drops to 0 and there is no positive gate to cathode voltage.
3. Conduct current in only one direction, anode to cathode, like a rectifier.
4. Be triggered by a gate current of only positive gate to cathode polarity.

Converters

Converters are classified by whether they convert ac to dc or vice versa; by whether single-phase or three-phase is used; by whether their output is variable; and by how many quadrants of a V versus I graph they operate in (see Figure 21-2).

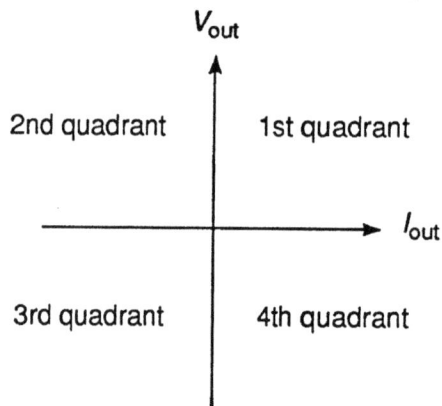

FIGURE 21-2 Converter Quadrants

The simplest converters convert electrical power in only the first quadrant. The converter of this experiment falls into this class. It can only convert ac power to dc power and will not convert dc back to ac.

A two-quadrant converter will convert ac power to dc power and convert dc power to ac power. A MG set (ac induction *M*otor powering a dc *G*enerator) is a two-quadrant converter.

For example, suppose a MG set is powering a dc motor that is operating a crane. If the crane is lowering a heavy load and the load overhauls its dc drive motor, that motor will turn into a generator and produce a higher than ordinary dc line voltage. This causes the MG set's generator to turn into a motor and send mechanical power to its ac drive motor. Then the ac drive motor also changes and becomes, for the moment, an induction generator and sends ac power back to the ac bus.

There are also four-quadrant converters. These are equivalent to two-quadrant converters with reversible output terminals.

The Converter Being Tested

The converter that has been constructed for this experiment is a simplified version of an industrial converter. Its industrial equivalent would have more sophisticated circuitry triggering its SCRs and might be a two- or four-quadrant converter, but would otherwise be very similar.

The converter being tested consists of six SCRs that are triggered by six matched RC circuits. The capacitor of each RC circuit charges during the positive half of its SCR's ac supply voltage. When the capacitor is sufficiently charged, its voltage breaks across its neon bulb to turn on its SCR. By adjusting the resistance in series with the capacitor, the amount of time it takes the capacitor to charge and turn on its SCR can be controlled. By controlling the amount of time it takes each SCR to turn on, voltage delivered by them to the dc output is controlled. This in turn controls the average converter output voltage. The commutation (turning off) of each SCR is accomplished during the negative half of each SCR's ac supply voltage, when the SCRs are reverse biased.

104

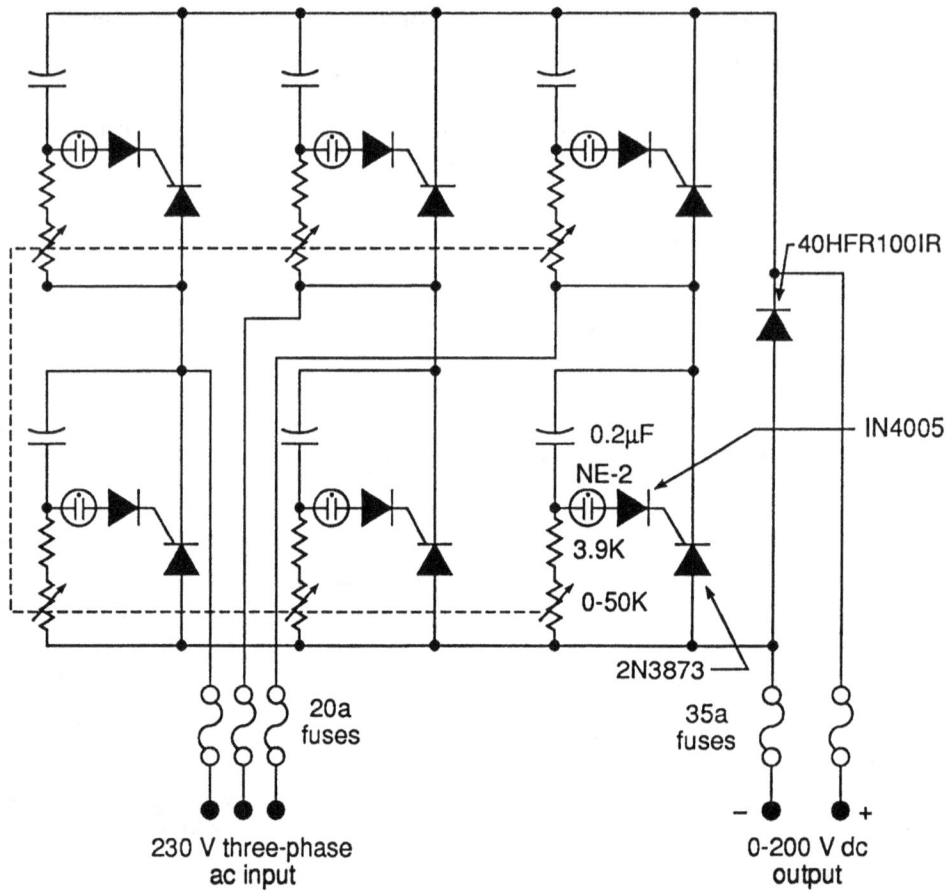

FIGURE 21-3 Three-Phase SCR Converter with RC Phase Shift Triggering

-PROCEDURE-

1. Construct the circuit of Figure 21-4 and have it approved by the instructor.

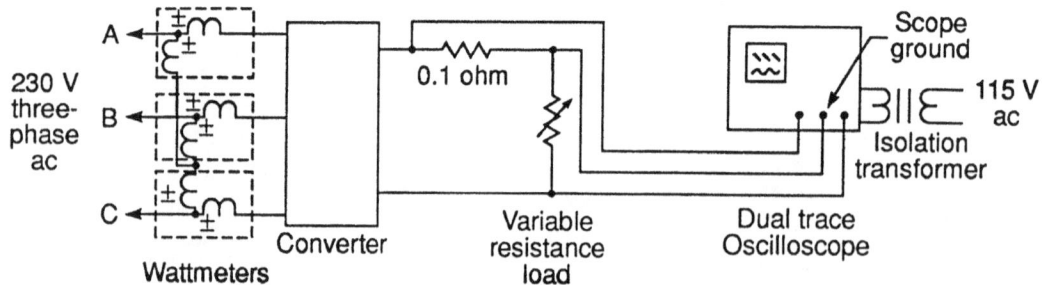

FIGURE 21-4

2. Set the variable resistance load to approximately 17 ohms.
3. Energize the converter. Adjust the converter's potentiometers until the power into the converter is 650 watts. Record the potentiometer setting, the voltages, and currents into the converter and accurately sketch the voltage and current waveforms of the converter's output. Repeat for 300 and 1000 watts of power into the converter.
4. Construct the circuit of Figure 21-5 and have it approved by the instructor.

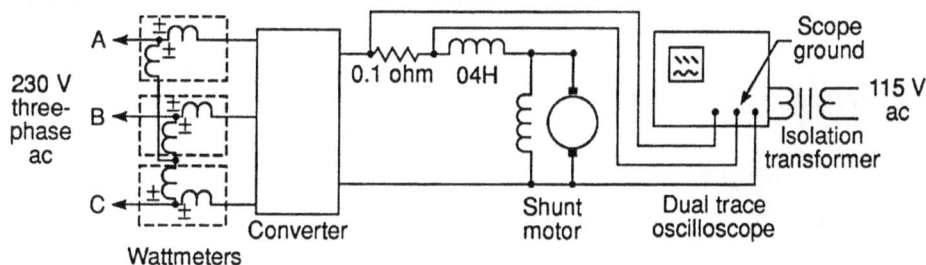

FIGURE 21-5

5. Set the converter's potentiometer setting to that of Procedure 3 when the input was 650 watts. Adjust the dynamometer load on the motor until the input power to the converter is 650 watts.
6. Take all the readings in Procedure 3.
7. Repeat Procedures 5 and 6 for 1000 watts input to the converter.

-QUESTIONS-

1. Determine the average output power of the converter for each load at each potentiometer setting. To do this, it will be necessary to find the average of the products of the instantaneous voltages times the instantaneous currents of the sketched waveforms.
2. On a single piece of graph paper, plot the efficiency and power factor of the SCR converter versus output power. Comment on the curves.
3. A Ward-Leonard system converter consists of a MG set and a low-power variable dc supply. The MG set's dc generator receives field voltage from the low-power variable dc supply. By controlling the low-power variable dc supply, the dc generator output voltage and power are controlled. What are some advantages and disadvantages of this experiment's converter compared to a Ward-Leonard system converter?

EXPERIMENT 22

OPERATION OF A SINGLE-PHASE MOTOR POWERED BY AN INVERTER

-OBJECTIVE-

To test the operation of a SCR, self-commutating, parallel capacitor commutated single-phase inverter as a speed controller for a single-phase induction motor.

-EQUIPMENT-

Inverter (already assembled)
4 H inductor
Single-phase induction motor
Fixed 45 volt dc supply
Variable frequency supply
Dynamometer
Ac ammeter
Ac voltmeter
2 dc ammeters
2 dc voltmeters
VOM
2 wattmeters
Oscilloscope
Isolation transformer
Tachometer
2PDT switch
2PST switch

-DISCUSSION-

Inverters change dc to ac. They vary in complexity from simple circuits, such as that of this experiment, to large, sophisticated, expensive electronic systems. For three-phase induction motor control, they produce three-phase within a frequency range of a few Hz to several hundred Hz. Inverters can produce other frequencies; however, induction motors usually don't operate well at lower or higher frequencies. Commercial inverters usually have a built-in method for dropping the inverter's output voltage in direct proportion to the output frequency to avoid saturating the motor's laminations at low frequency. Finally, commercial inverters are often connected into a speed control system with feedback from a tachometer and external signals.

In the past, inverter/three-phase induction motor control systems were more expensive than equivalently capable dc motor control systems. For many applications this is no longer true. In addition, the inverter/three-phase induction motor system has an advantage in that three-phase induction motors don't have brushes and a commutator to spark and wear out as dc motors do. In applications where sparking is not allowed, such as in mines, or where the motor is not easily accessible for servicing, inverter/three-phase induction motor systems should be used.

Single-Phase SCR Self-Commutating Parallel Capacitor Commutated Inverter

A simple single-phase self-commutated parallel capacitor commutated inverter is shown in Figure 22-1. Sketches of its voltage and current waveforms are shown in Figure 22-2. The SCRs of this inverter alternately turn on, as each receives a pulse from the triggering circuit. Consider an instant when SCR 1 is conducting and SCR 2 is not; call this period 1. During period 1, the bottom side of the capacitor is charging positively while the top side is effectively grounded. At the start of period 2, SCR 2 is triggered. At this instant both SCRs are on. Now the charge that is stored on the bottom side of the capacitor will rapidly start to discharge clockwise through SCR 1. When the current through SCR 1 goes to zero, SCR 1 turns off. Now the cycle of events repeats with SCR 1 off and SCR 2 on. The SCRs' currents, cycling through the transformer's primary, induce the output voltage on the transformer's secondary.

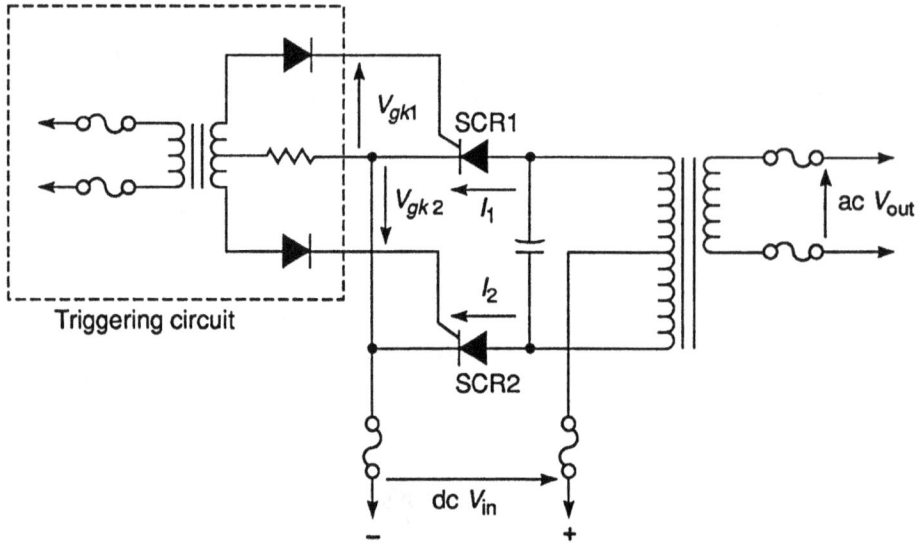

FIGURE 22-1 Simple Inverter

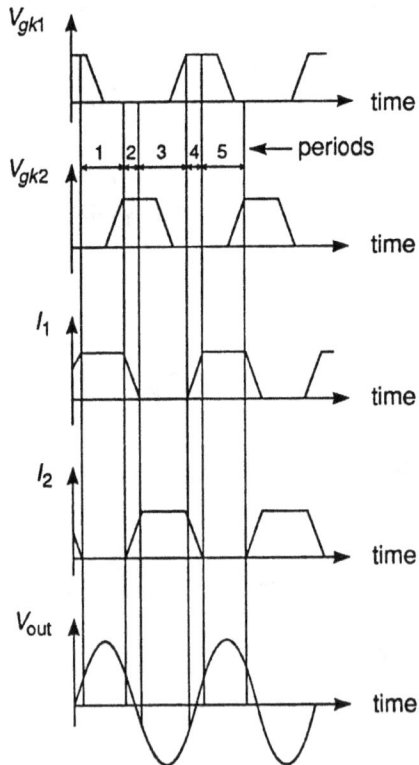

FIGURE 22-2 Simple Inverter Waveforms

Inverter Being Tested

The inverter that has been constructed for this experiment is very similar to that described above. There are only two differences. First, the commutating capacitor is located on a high-voltage secondary winding of the transformer. This effectively multiplies the capacitance by the square of the transformer's turns ratio. The 15 μF capacitor in the circuit would be worth 15 x $(440/180)^2$ = 90 μF as the capacitor of Figure 22-1. The higher capacitance represents an advantage, for it gives the inverter a higher current capacity and more ability to handle an inductive load.

Second, snubbers are connected from the anode to cathode of each SCR. These snubbers are resistors in series with capacitors. The capacitors of the snubbers reduce the voltage spikes and high dv/dt that the circuit inductance produces across the SCRs. This reduction is necessary to avoid falsely triggering the SCRs.

The quality of dc used is important to this inverter. This inverter will operate properly when it is supplied dc from a generator or full-wave rectified three-phase through a 4 H inductor. Single-phase full-wave rectified dc has too much ripple. Too much ripple causes the SCRs to falsely trigger.

The inverter circuit used is shown in Figure 22-3. The ac voltages shown on the inverter, 100, 180, and 440, were measured when the inverter was supplying 60 Hz to a 1/4 single-phase induction motor connected to an unloaded dynamometer. The voltages in this inverter are frequency and load dependent. Other voltage magnitudes will be observed under other load conditions.

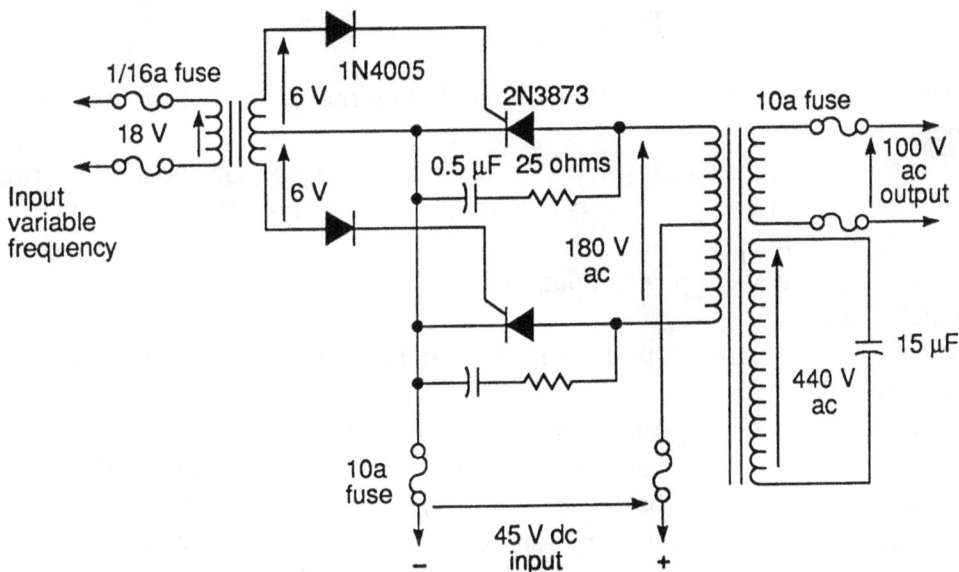

FIGURE 22-3 Inverter Being Tested

-PROCEDURE-

1. Construct the circuit of Figure 22-4 and have it approved by the instructor.

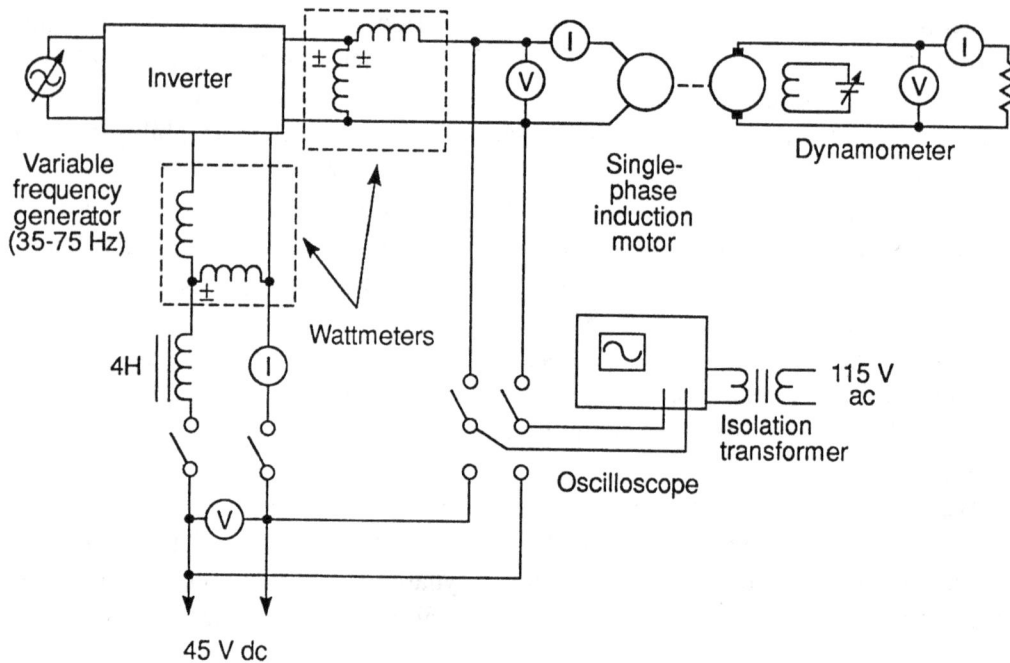

FIGURE 22-4

2. Using the oscilloscope, sketch the waveform of the dc power supply's voltage before the inverter is energized.
3. Turn on the variable frequency supply. Adjust its output to supply 55 Hz and 12 volts rms at the triggering transformer's output (6 volts rms from the center tap to an outer terminal).
4. Apply 0 volts to the dynamometer's field.
5. Apply 45 volts dc to the inverter's power input.
6. Measure the motor's rpm.
7. Record input dc volts, amps, and watts, as well as inverter output ac volts, amps, and watts as seen on the meters.
8. Sketch the input and output voltage waveforms.
9. Repeat Procedures 3 to 8 with the variable frequency supply set to 35 Hz. At the lower motor speed that will be produced, be certain to disconnect the starting winding from the circuit after starting. At lower operating speeds, the centrifugal switch might not disconnect, which would cause the starting winding to continue to receive voltage and overheat.

10. Repeat Procedures 3 to 8 with the variable frequency supply set to 45, 65, and 75 Hz.
11. Return the variable frequency supply to 55 Hz.
12. Do a load test on the inverter.
   a. Determine the range of power outputs available from the inverter by turning up the dynamometer's field voltage until the motor speed starts to suddenly decrease.
   b. Divide the dynamometer's maximum field voltage into five increments.
   c. Run the inverter with the load produced by the dynamometer at each field increment. Take all the readings of Procedure 7 and the dynamometer output voltage and current at each increment.

-QUESTIONS-

1. In your own words explain how this inverter works.
2. Draw the voltage waveforms.
3. How did the waveforms change with frequency?
4. Graph the output voltage of the inverter versus frequency for the circuits of Procedures 3 to 10.
5. Graph the inverter's efficiency versus frequency for the circuits of Procedures 3 to 10.
6. Graph the inverter's efficiency versus inverter power out for the load test of Procedure 12.
7. Where is power lost in the inverter?

INTRODUCTION TO PROGRAMMABLE CONTROLLERS EXERCISE

-OBJECTIVE-

To become familiar with programmable controllers.

-MATERIALS-

Sales brochures. Contact the following manufacturers or their local representatives to request free programmable controller sales brochures.

Allen-Bradley, division
of Rockwell Automation
1201 South Second Street
Milwaukee, WI 53204
1-414-382-2000
http://www.ab.com/

GE Fanuc Automation
North America, Inc.
Information Center
P.O. Box 4248
Lynchburg, VA 24502
1-800-648-2001
http://www.geindustrial.com/cwc/gefanuc/index.jsp

Siemens Energy & Automation, Inc.
3333 Old Milton Parkway
Alpharetta, GA 30202
1-800-964-4114
http://www.sea.siemens.com/

-DISCUSSION-

To most people a programmable controller, programmable logic controller and PLC are the same thing. Although it should be noted that PLC is registered trademark that the Allen Bradley company uses to describe one of its lines of programmable controllers.

Programmable controllers are simply special purpose computers that control electrically operated processes. The processes might be in chemical plants, steel mills or any other type of industrial plant that needs precise control. Often programmable controllers do jobs that were formerly done by teams of human operators.

Any process controlled by a programmable controller has the following:
1. The process being controlled.
2. Input devices such as switches, sensors, or pushbuttons.
3. Input modules that act as a protective boundary and convert signals from the input devices to a form that can be used by the programmable controller's central processing unit, communication, and memory.
4. The programmable controller's central processing unit (CPU), communication and memory and power supply.
5. A software program.
6. Output modules that act as a protective boundary and convert the CPU's output to a form that can operate external devices.
7. External devices such as lights, solenoids, and motor starters.
8. Operator terminal for programming and monitoring the control system and the process.

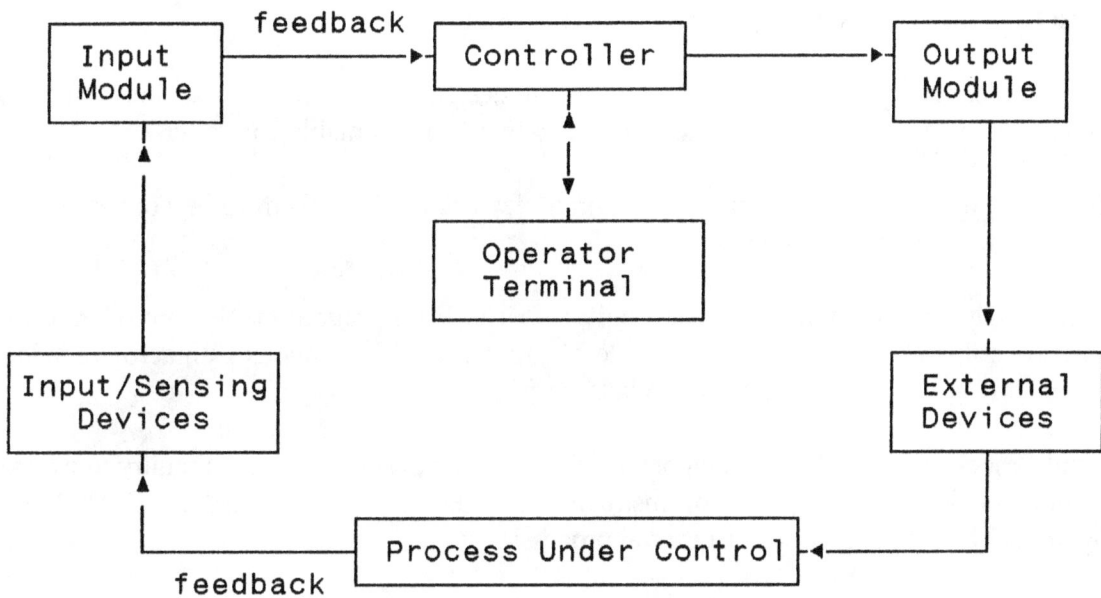

FIGURE 23-1 Typical programmable controller configuration.

There are few programmable controller construction standards and there are a great number of manufacturers. Furthermore, the designs and programming languages of the programmable controllers of each manufacturer are rapidly evolving. For these reasons it is difficult for a school to offer hands-on laboratory exercises on up-to-date programmable controllers. Most colleges and universities have outdated programmable controllers or none at all.

114

If your school has old programmable controllers, use them. The old programmable controllers are similar enough to the new that it is worth getting some hands-on experience with them. If at sometime in the future you work with more modern programmable controllers you will then just update your skills with on-the-job learning or training from the manufacturer.

Programmable controller systems have replaced relay logic systems in most new installations. They are more flexible, more rugged, easier to reprogram, and less expensive than all but the simplest relay logic systems. Also, programmable controllers can receive more types of inputs than relays. For example, some can receive input data from bar code scanners. Programmable controller knowledge is of use to maintenance technicians, plant engineering personnel, technologists, and engineers.

PROGRAMMABLE CONTROLLER GLOSSARY

Address: Identifying numbers and/or letters that designate a particular I/O location on a programmable controller or on a device controlled by a programmable controller.

Adapter Board: Module that is used on external devices to allow them to be connected to a programmable controller's data bus.

Bit: Generally speaking this is a binary digit, a 1 or 0. With programmable controllers using ladder logic this is considered an "on" or "off". An "on" is a continuous path across a ladder rung an "off" is a break in the circuit of a ladder rung.

Central Processing Unit: Major component of a computer system with the circuitry to control the interpretation and execution of instructions. Often it is abbreviated as CPU. Every programmable controller contains a central processing unit.

EEPROM: Electrically erasable programmable read-only memory.

Firmware: Programmable controller instructions that are embedded in the hardware, stored in PROM, EPROM or EEPROM devices, and are generally not modifiable by the user.

Instruction Set or Instructions: These are the program statements used to manipulate data received by the programmable controller. Some of the instructions available on programmable controllers are: relay-type ("on" and "off"), timer, counter, comparison, arithmetic, Boolean logic, and program control.

I/O: Abbreviation for "Input/Output". A programmable controller receives data through its input terminals and sends out signals and controlling voltages through its output terminals.

Ladder Diagram: Standard method of drawing relay logic control circuits. The drawings resemble a ladder. Examples can be seen in Figures 18-2 and 19-1 on pages 89 and 92. Most programmable controller manufacturers use software created ladder diagrams as part of their programming languages.

Module: Interchangeable plug-in electronic item, often a printed circuit board or card.

Open Architecture: Computer or programmable controller design for which detailed specifications are published by the manufacturer, allowing others to produce compatible hardware and software.

Port: Connector or terminal strip used to access a system of circuit. Usually ports are used for the connection of peripheral equipment.

Programming Terminal: Keyboard device used to input programs and data and operate a programmable controller. It can be mounted on the programmable controller, a separate hand-held device, or a specially configured personal computer.

Rack: Framework or chassis that houses programmable controller modules. From a programming prospective, a single programmable controller framework or chassis sometimes contains more than one rack.

Register: Data storage location in a computer or programmable controller.

Retentive Register: Data storage location that retains its data during a power down.

Scan Time: The time to read all inputs, execute the control program, and update all input and output statuses. It is the time required to activate an output that is being controlled by a programmable controller. If the scan time is too long, a programmable controller may not be able to successfully control a process.

Statement Language: Sometimes abbreviated as STL. Programming language for programmable controllers that uses statements like those found in the BASIC computing language rather than ladder diagrams.

Upward Migration: Term that indicates that a programmable controller can do every thing that its previous model could do, plus some additional functions.

116

-QUESTIONS-

1. Select one programmable controller sales catalog and write an outline of it. (Make the outline about two handwritten pages long.)

2. In your selected sales catalog, what are some of the features common to all the programmable controllers.

3. In your selected sales catalog, which programmable controllers can be configured with the most I/O terminals.

4. How fast can some of the programmable controllers of your selected sales catalog execute a command after receiving a signal input. In your answer refer to the location in the catalog where you got this number.

5. Using a technical dictionary and sales catalogs, define the following words in relation to programmable controllers: analog I/O module, discrete I/O module, family, flexibility, modular, programming software, and system configuration.

APPENDIX A

ACCURACY

INSTRUMENT ACCURACY

No instrument is perfect; there is always a limit to its accuracy.

The instrumentation used in an electrical machinery laboratory may have a percentage error as high as ± 5 percent. To the experimenter, that means recording data to three significant figures is sufficient. Taking readings with more than three significant figures will not invalidate the experimenter's data but may waste time and lead to a false impression of the worth of the data and the conclusions that may be drawn from them.

ROUND-OFF ERRORS IN CALCULATIONS

Calculations may induce errors resulting from improper rounding off of numbers. An example of round-off error is shown where all calculations are rounded off to three significant figures. It is contrasted with the same calculation done with all calculations rounded off to 10 significant figures.

R = 1.34 ohms, L = 0.00319 H, C = 0.00101 F and w = 377 rad/sec
$$Z = \{R^2 + [wL - 1/(wC)]^2\}^{.5}$$

With the calculations carried out to three significant figures, the following is obtained:

$$Z = \{1.34^2 + [377 \times 0.00319 - 1/(377 \times 0.00101)]^2\}^{.5}$$
$$Z = \{1.80 + [1.20 - 1/0.381]^2\}^{.5}$$
$$Z = \{1.80 + [1.20 - 2.62]^2\}^{.5}$$
$$Z = \{1.80 + [-1.42]^2\}^{.5}$$
$$Z = \{1.80 + 2.02\}^{.5}$$
$$Z = \{3.82\}^{.5}$$
$$Z = 1.95 \text{ ohms}$$

With the calculations carried out to 10 significant figures throughout, the following is obtained.

$$Z = \{1.340000000^2 + [377.0000000 \times 0.003190000000 - 1/$$
$$(377.0000000 \times 0.001010000000)]^2\}^{.5}$$
$$Z = \{1.795600000 + [1.202630000 - 1/.3807700000]^2\}^{.5}$$
$$Z = \{1.795600000 + [1.202630000 - 2.626257321]^2\}^{.5}$$
$$Z = \{1.795600000 + [-1.423627321]^2\}^{.5}$$
$$Z = \{1.795600000 + 2.026714748\}^{.5}$$
$$Z = \{3.822314748\}^{.5}$$
$$Z = 1.955074103 \text{ ohms}$$

Properly rounded off to three figures in the final answer, $Z = 1.96$ ohms.

The answer determined with all calculation numbers rounded off to three significant figures is in error by 0.01 ohms or 0.5 percent. The error was caused by too much rounding off. Further calculations with the incorrect Z could create even bigger errors.

For practical laboratory work, calculations using data with three significant figures of accuracy should be carried out with at least four significant figures. Once the calculations are complete, the final answer should be expressed as no more than three significant figures.

The calculations of the example carried out to four significant figures give $Z = 1.956$, which then properly rounds off to $Z = 1.96$ ohms.

Percentage Error and Percentage Difference

It is often informative to calculate the percentage error of a measuring instrument or the percentage difference between two measurements.

For example, suppose a circuit voltage is known to be 5.00 volts and is measured experimentally to be 5.20 volts. The percentage error expresses the inaccuracy of the measuring instrument.

% error = [(Measured value - actual value)/actual value] x 100%

% error = [(5.200 - 5.000)/5.000] x 100% = 4.00 %

Measuring instruments are usually rated with a $\pm$ percentage error, which shows the range of possible actual values indicated by an instrument. For example, suppose a voltmeter that is rated at $\pm$ 3 percent error indicates 100 volts. The actual voltage it is measuring is some unknown value between 100 - (0.03)(100) = 97 volts and 100 + (0.03)(100) = 103 volts. Knowledge of the range of possible actual values of a measured value can be useful when drawing conclusions.

Percentage difference is calculated in a similar way to percentage error.

% difference = [(1st value - 2nd value)/1st value] x 100%

APPENDIX B

EQUIPMENT

VOLTAGE SUPPLIES

Voltage supplies are listed as fixed or variable. Fixed ac supplies are simply connections to the ac bus. Variable ac supplies are produced by variacs connected to the ac bus. Fixed dc supplies are produced by bridge rectifiers connected to the ac bus. Variable dc supplies are produced by bridge rectifiers connected to variacs that are connected to the ac bus. Usually, voltages are specified as 115 volts. This number is only a guide; the equipment being tested may require a different voltage.

Each **EQUIPMENT** list gives the number of voltage supplies. This number is the maximum that should be necessary for the experiment. It will sometimes be possible to reduce the number of supplies in the actual experiment. For example, a dynamometer's generator is assumed to be a separately-excited generator and therefore requires a dc voltage supply. With some experiments, the generator may be self-excited, eliminating the need for a variable dc power supply. Likewise, each experiment using a dynamometer as a drive motor is assumed to have two variable dc supplies, one for the field and one for the armature. Although it might make the experiment more cumbersome, it is sometimes possible to reduce that number to one fixed supply with a rheostat in series with the field winding.

DYNAMOMETER

Dynamometers are the best machines for measuring torque supplied to or received from the rotating shaft of a machine or motor. They are specially mounted dc motors or generators, depending on how they are used. A dynamometer's stator is mounted on bearings so that it can attempt to rotate in the direction the armature magnetic field pushes or pulls it. Connected to one side of the stator is a scale that stops the stator rotation and measures the rotational torque on the stator. For a more detailed description, see an electrical machinery text.

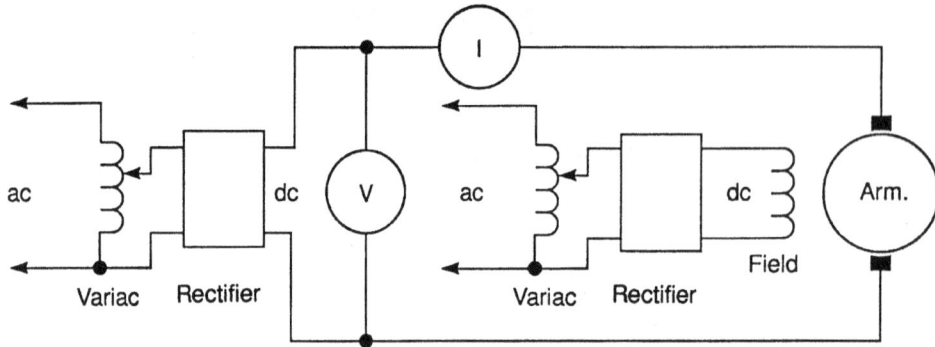

FIGURE B-1 Recommended Connections for a Dynamometer as a Drive Motor

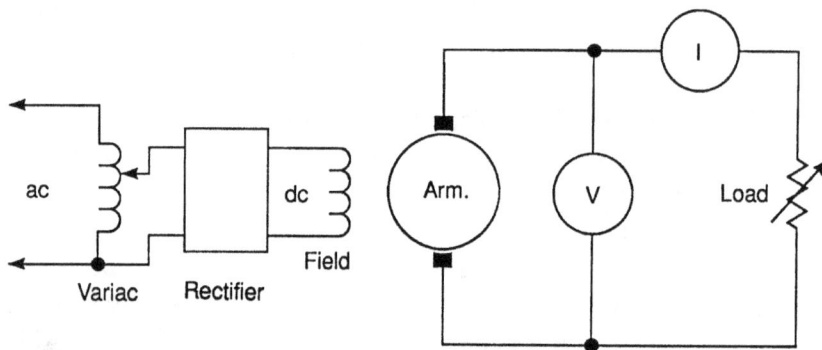

FIGURE B-2 Recommended Connections for a Dynamometer as Generator

CALIBRATED DC MOTOR AND CALIBRATED DC GENERATOR

It is possible to use a calibrated dc motor or generator instead of a dynamometer.

A separately-excited dc motor with a fixed voltage applied to its field will produce a certain torque for each current supplied to its armature.

A separately-excited dc generator connected to a fixed load will use a certain amount of torque for each voltage supplied to its field at each rpm.

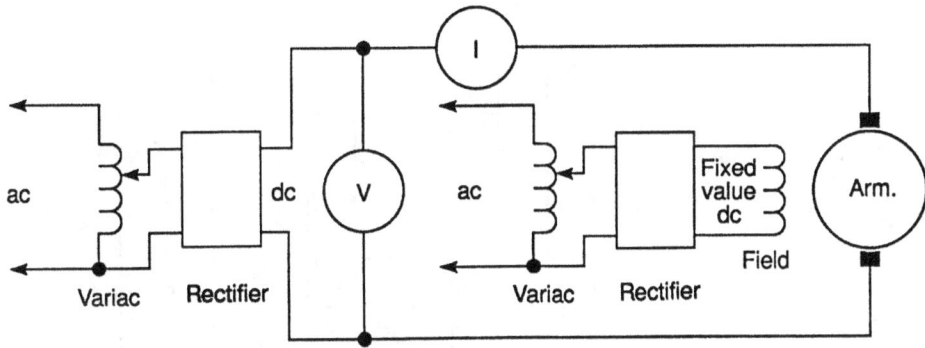

FIGURE B-3 Connection Diagram for a Calibrated Dc Motor

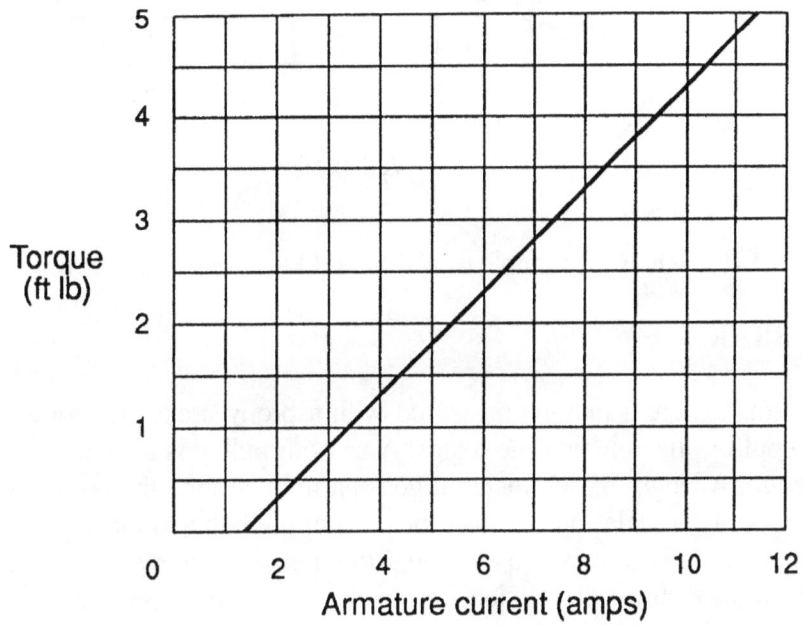

FIGURE B-4 Example of a Calibrated Dc Motor Curve

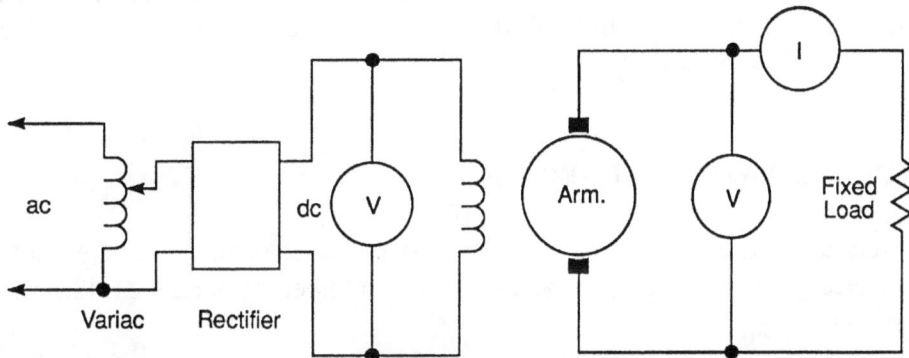

FIGURE B-5 Connection Diagram for a Calibrated Dc Generator

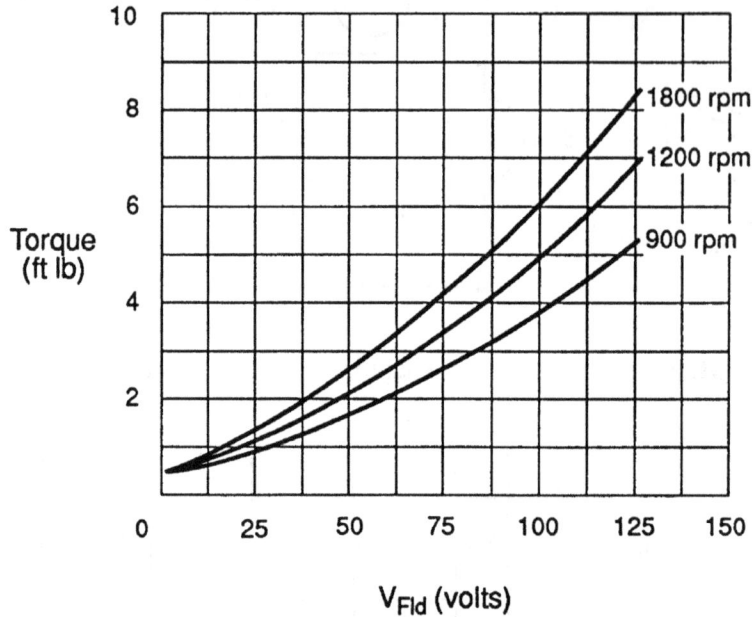

FIGURE B-6 Example of Calibrated Dc Generator Curves

PRONY BRAKE

Torque from a rotating shaft may be measured with a prony brake. A prony brake has three main parts: the pulley, the belt, and the scales. A smooth pulley is attached to the shaft of the motor being tested. A length of belt material goes partially around the pulley and rubs against the pulley's surface. Two scales pull against the ends of the belt and thereby stop the belt from rotating with the pulley. Torque is equal to the difference of scale readings times the radius of the pulley. For a more detailed description, see an electrical machinery text.

A prony brake is simpler and cheaper than the previously mentioned electrical methods of torque measurement. However, a prony brake has several disadvantages. First, it will not measure torque supplied to a load. Second, the operation of a prony brake is not smooth. There is a tendency for the belt to suddenly stall the motor, especially at low rpm. Third, heating of the belt may be a problem for longer tests.

INSTRUMENTATION (GENERAL)

Enough meters are specified to measure all important variables and to make the moving of meters unnecessary during testing. If necessary, the number of meters could be reduced in some experiments.

AC CLAMP-TYPE AMMETER

Ac clamp-type ammeters have been specified in many experiments because they are often easier and faster to use than in-line ac ammeters.

However, clamp-type ammeters have two practical problems.

First, they often aren't as accurate as in-line meters. This is not a problem with the experiments of this book, which do not require higher accuracy than a clamp-type ammeter can provide.

Second, they don't have small current ranges. This problem can be solved by putting loops of the current carrying conductor through the meter's clamp. For example, suppose a conductor carrying 1 amp is to have its current measured by a clamp-type ammeter with a minimum full-scale current of 6 amps. Of course, the clamp could be placed around just one conductor and the meter would read 1 amp. That might be satisfactory, but if a more accurate reading were desired, the conductor could be looped twice through the clamp. Looping the conductor twice would double the sensitivity of the meter. Now the meter would indicate 2 amps on its scale. The experimenter would divide the 2 amps by 2 to get the actual conductor current. This can be expressed in the general equation:

$$I = I_M/N$$

I = actual conductor current

I_M = current read on the meter

N = Number of conductor loops through the clamp

VARIABLE RESISTANCE LOADS

Usually, light bulbs are used as loads in these experiments.

APPENDIX C

MOTOR RUNAWAY

Dc motors and ac series motors may suddenly over speed under certain conditions. The over speeding is called runaway. It may damage the motor, the motor's load, or people near the motor. Runaway is dangerous and should be avoided.

DC SHUNT MOTOR RUNAWAY

A properly operating unloaded dc shunt motor may be prone to runaway if its shunt field is disconnected. The reason for this can best be seen with the motor equations:

$$V_I = E_G + I_A \times R_A$$

$$E_G = K \times \varphi \times S$$

where

V_I = voltage applied to the motor armature, a constant
E_G = internally generated armature voltage
I_A = current going to the motor armature
R_A = resistance of the motor armature, a constant
φ = flux produced by the shunt field
S = speed of the motor (rpm)
K = a constant that depends on the motor construction

The equations can be combined.

$$V_I = K \times \varphi \times S + I_A \times R_A$$

This equation can be solved for S.

$$S = (V_I - I_A \times R_A)/(K \times \varphi)$$

In the runaway condition, the shunt field winding is not producing field, so φ is very small. The very small φ on the denominator causes S to become very large; as a result, runaway occurs.

DC OR AC SERIES MOTOR RUNAWAY

Series motors can run away without having their series field disconnected. The field φ, of a series motor, is dependent on the armature current. The armature current depends on its shaft load. If the shaft load on a series motor's output is decreased, its armature current will decrease, causing φ to decrease. Using the same equation as used with the dc shunt motor, it can be seen that an unloaded series motor is prone to runaway.

$$S = (V_I - I_A \times R_A)/(K \times \varphi)$$

To avoid series motor runaway, be certain to always to have a shaft load on series motors.

DC COMPOUND MOTOR RUNAWAY

If an unloaded compound motor's shunt field is disconnected, it will run away in the same way as an unloaded series motor.

REFERENCES

Emanuel, P. 1990. *Motors, Generators, Transformers, and Energy*. Englewood Cliffs, N.J.: Prentice-Hall.

Fink, D. G., and Beaty, H. W.(eds.). 1999. *Standard Handbook for Electrical Engineers*. New York: McGraw-Hill.

Richardson, D. V., and Caisse, A. J. 1996. *Rotating Electric Machinery and Transformer Technology*. Reston, Va.: Prentice Hall.

Rosenblatt, J., and Friedman, M. H. 1998. *Direct and Alternating Current Machinery*. Columbus, Ohio: Merrill.

Wildi, T. 2001. *Electrical Machines, Drives, and Power Systems*. Englewood Cliffs, N.J.: Prentice Hall.

www.ingramcontent.com/pod-product-compliance
Lightning Source LLC
Chambersburg PA
CBHW080600220326
41599CB00032B/6550